STANDARD AND
NON-STANDARD METHODS
FOR SOLVING ELEMENTARY
ALGEBRA PROBLEMS

Essential Textbooks in Mathematics

ISSN: 2059-7657

The *Essential Textbooks in Mathematics* series explores the most important topics that undergraduate students in Pure and Applied Mathematics are expected to be familiar with.

Written by senior academics as well lecturers recognised for their teaching skills, they offer, in around 200 to 400 pages, a precise, introductory approach to advanced mathematical theories and concepts in pure and applied subjects (e.g. Probability Theory, Statistics, Computational Methods, etc.).

Their lively style, focused scope, and pedagogical material make them ideal learning tools at a very affordable price.

Published:

Standard and Non-Standard Methods for Solving Elementary Algebra Problems
　　by V G Chirskii (Lomonosov Moscow State University, Russia & RANEPA, Russia) and A I Kozko (Lomonosov Moscow State University, Russia & RANEPA, Russia)

Ordinary Differential Equations and Applications
　　by Enrique Fernández-Cara (University of Seville, Spain)

A First Course in Algebraic Geometry and Algebraic Varieties
　　by Flaminio Flamini (University of Rome "Tor Vergata", Italy)

Analysis in Euclidean Space
　　by Joaquim Bruna (Universitat Autònoma de Barcelona, Spain & Barcelona Graduate School of Mathematics, Spain)

Introduction to Number Theory
　　by Richard Michael Hill (University College London, UK)

A Friendly Approach to Functional Analysis
　　by Amol Sasane (London School of Economics, UK)

A Sequential Introduction to Real Analysis
　　by J M Speight (University of Leeds, UK)

Essential Textbooks in Mathematics

STANDARD AND NON-STANDARD METHODS FOR SOLVING ELEMENTARY ALGEBRA PROBLEMS

V G Chirskii

A I Kozko

Lomonosov Moscow State University, Russia & RANEPA, Russia

World Scientific

NEW JERSEY · LONDON · SINGAPORE · BEIJING · SHANGHAI · HONG KONG · TAIPEI · CHENNAI

Published by

World Scientific Publishing Europe Ltd.

57 Shelton Street, Covent Garden, London WC2H 9HE

Head office: 5 Toh Tuck Link, Singapore 596224

USA office: 27 Warren Street, Suite 401-402, Hackensack, NJ 07601

Library of Congress Cataloging-in-Publication Data

Names: Chirskii, V. G., author. | Kozko, A. I., author.
Title: Standard and non-standard methods for solving elementary algebra problems /
    V.G. Chirskii, A.I. Kozko, Lomonosov Moscow State University, Russia, RANEPA, Russia.
Description: New Jersey : World Scientific, 2025. | Series: Essential textbooks in mathematics,
    2059-7657 | Includes bibliographical references and index.
Identifiers: LCCN 2024007318 | ISBN 9781800615717 (hardcover) |
    ISBN 9781800615861 (paperback) | ISBN 9781800615724 (ebook for institutions) |
    ISBN 9781800615731 (ebook for individuals)
Subjects: LCSH: Algebra--Textbooks.
Classification: LCC QA152.3 .C45 2025 | DDC 512.9--dc23/eng/20240412
LC record available at https://lccn.loc.gov/2024007318

British Library Cataloguing-in-Publication Data
A catalogue record for this book is available from the British Library.

For any available supplementary material, please visit
https://www.worldscientific.com/worldscibooks/10.1142/Q0464#t=suppl

Desk Editors: Balasubramanian Shanmugam/Rosie Williamson/Shi Ying Koe

Typeset by Stallion Press
Email: enquiries@stallionpress.com

Preface

This book contains an exposition of methods for solving problems related to elementary algebra, both simple and quite difficult. For the convenience of readers, it contains all the basic information related to the methods of solving equations and inequalities. The basic properties of elementary functions are also described, and useful inequalities are given. Separately, the chapters contain problems related to rational equations and inequalities, to equations and inequalities with roots, and to equations and inequalities containing the sign of the absolute value. Equations and inequalities with exponential and logarithmic functions and trigonometric equations and inequalities are considered, as well as equations and inequalities of mixed type. Furthermore, we consider systems of equations, the methods of their solutions, and problems with integers.

Some important topics of analysis are included, for example, the Jensen and Cauchy inequalities and the concepts of concavity.

Special attention is paid to solving problems with parameters. As a rule, they cause great difficulties. At the same time, the logical conclusions that need to be drawn in the decision process provide excellent training for mental abilities. The book discusses various approaches to their solutions. Moreover, we obtain solutions to some difficult tasks based on the use of different techniques. This allows the interested reader to choose the most suitable ways to solve problems for themselves. For example, the area method, which is similar to the interval method for inequalities with one variable. Another method involves reversing the variable and parameter, and the problem is solved relative to the parameter. In many cases, the use of suitable groupings and variable replacement is required. Examples are the method of using some inequalities and the tasks with iterations. We adopt geometric approaches to solving algebraic problems and a number of other methods. The book contains a large number of tasks as exercises for readers. These tasks mainly appeared at entrance exams and various Olympiads.

The book can be useful for schoolchildren, math teachers, and readers interested in mathematics.

V. G. Chirskii
A. I. Kozko

About the Authors

Vladimir Grigoryevich Chirskii graduated from Lomonosov Moscow State University in 1972 and enrolled in the postgraduate course of the Faculty of Mechanics and Mathematics at the same university. His supervisor was the famous scientist Professor A. B. Shidlovskii, one of the creators of the Siegel–Shidlovskii method in the theory of transcendental numbers. From 1975 to the present, V. G. Chirskii has been primarily working at the Faculty of Mechanics and Mathematics, Lomonosov Moscow State University. He is a doctor of physical and mathematical sciences (equivalent to doctor habilitatus) and is currently a professor at the Department of Mathematical Analysis, Lomonosov Moscow State University. The area of V. G. Chirskii's scientific research involves the theory of transcendental numbers in direct products of fields of p-adic numbers. This research focus began to develop with his work in the early 1990s. He has published over a hundred scientific papers on number theory and the applications of mathematical methods to problems in economics and more than twenty books, including those written for schoolchildren and teachers. Among his students are several candidates pursuing physical and mathematical sciences (equivalent to a Ph.D.) and one doctor of physical and mathematical sciences (equivalent to a doctor habilitatus). In addition to his role at Lomonosov Moscow State University, V. G. Chirskii headed the Department of Number Theory at the Moscow State Pedagogical University for several years. He also offers lectures at the Russian Academy of National Economy and Public Administration.

Artem Ivanovich Kozko graduated from the Faculty of Mechanics and Mathematics at Lomonosov Moscow State University. Since 1997, he has been teaching and lecturing at the university in the Department of Mathematical Analysis, as well as in the Faculties of Mechanics and Mathematics, Chemistry, and Psychology. Since 2013, he has been the senior examiner for the Olympiad "Conquer Vorobyovy Gory" at Lomonosov Moscow State University in mathematics. The experience of these exams is also reflected in this book.

A. I. Kozko's research interests relate to extreme problems in the theory of approximations, spectral problems in the theory of differential equations, as well as to economic models of growth in the economy. He has published more than 75 scientific articles and more than 20 books.

Contents

Chapter 1

Exponential and Logarithmic Equations and Inequalities

1.1 Basic Properties of Exponential and Logarithmic Functions

The function a^x is defined for $a > 0$ and for all x. It increases if $a > 1$ in Fig. 1.1 and decreases if $0 < a < 1$ in Fig. 1.2. The set of its values is the set of real numbers. For all $a > 0$, $b > 0$, x, and y, one has:

(1) $a^x a^y = a^{x+y}$,
(2) $a^x / a^y = a^{x-y}$,
(3) $a^x b^x = (ab)^x$,
(4) $a^x / b^x = (a/b)^x$,
(5) $(a^x)^y = a^{xy}$.

The logarithmic function $\log_a x$ is defined for $a > 0$, $a \neq 1$, and $x > 0$. The equality $y = \log_a x$ is equivalent to the equality $a^y = x$. The logarithmic function increasesin Fig. 1.3 if $a > 1$ and decreases if $0 < a < 1$ in Fig. 1.4. The set of its values is the set of real numbers. For all $a > 0$, $a \neq 1$, $x > 0$, and $y > 0$, the following equalities hold:

(1) $a^{\log_a x} = x$;
(2) $\log_a x + \log_a y = \log_a xy$;
(3) $\log_a x - \log_a y = \log_a \frac{x}{y}$;
(4) $\log_a x^\alpha = \alpha \log_a x$;
(5) $\log_{a^\beta} x = \frac{1}{\beta} \log_a x$, $\beta \neq 0$;

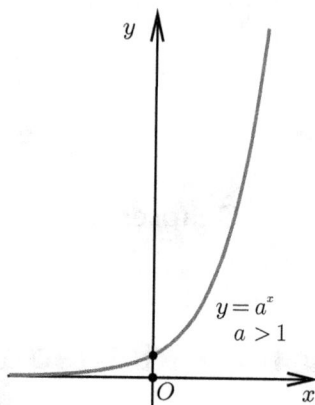

Fig. 1.1.   Graph of the function $y = a^x$, $a > 1$.

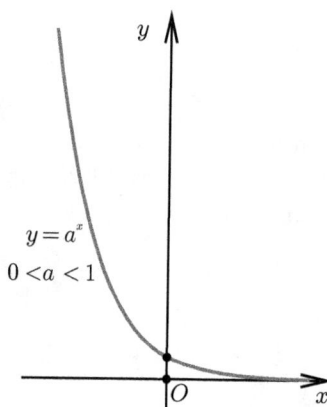

Fig. 1.2.   Graph of the function $y = a^x$, $0 < a < 1$.

(6) For all $a > 0$, $a \neq 1, c > 0$, $c \neq 1, b > 0$ $\log_a b = \frac{\log_c b}{\log_c a}$; in particular, when $a > 0$, $a \neq 1, b > 0$, $b \neq 1$ $\log_a b = \frac{1}{\log_b a}$.

It is useful to note a few more formulas:

(7) If $a > 0, a \neq 1, xy > 0$, then $\log_a xy = \log_a |x| + \log_a |y|$, $\log_a \frac{x}{y} = \log_a |x| - \log_a |y|$.

(8) If $x \neq 0, m \in \mathbb{N}$, then $\log_a x^{2m} = 2m \log_a |x|$.

(9) If $a > 0, b > 0, b \neq 1, c > 0$, then $a^{\log_b c} = c^{\log_b a}$.

(10) If $a > 0, b > 0, c > 0, a, b \neq 1$, then $\log_a b \log_b c = \log_a c$.

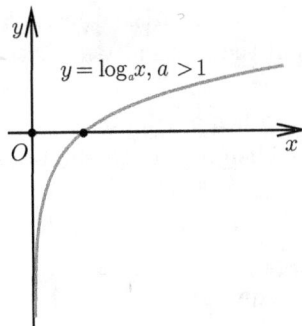

Fig. 1.3.   Graph of the function $y = \log_a x$, $a > 1$.

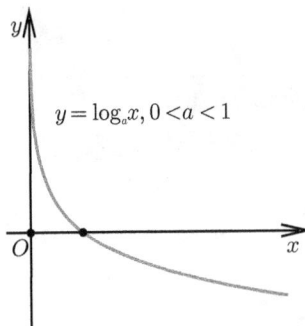

Fig. 1.4.   Graph of the function $y = \log_a x$, $0 < a < 1$.

Equality (9) follows from the fact that $\log_b a^{\log_b c} = \log_b a \log_b c = \log_b c^{\log_b a}$. Equality (10) is easy to get if you use formula (6) in the second logarithm.

The notation $\ln x$ is often used to denote the logarithm based on e (where $e > 1$ is a number equal to $\lim_{n \to \infty} \left(1 + \frac{1}{n}\right)^n$, designated in honor of the great L. Euler), and $\lg x = \log_{10} x$.

1.2   The Simplest Equations and Inequalities Containing Exponential and Logarithmic Functions

The equation $a^x = b$ has no solutions if $b \leq 0$. If $b > 0$, then this solution has the form $x = \log_a b$. The inequality $a^x < b$ has no solutions if $b \leq 0$. If $b > 0$ and $a > 1$, then the solutions of the inequality are $x < \log_a b$, and if $b > 0$ and $0 < a < 1$, then the solutions of the inequality are $x > \log_a b$.

The inequality $a^x > b$, if $b \leq 0$, is satisfied by all values of x. If $b > 0$ and $a > 1$, then the solutions of the inequality are $x > \log_a b$, and if $b > 0$ and $0 < a < 1$, then the solutions of the inequality are $x < \log_a b$.

The equation $\log_a x = b$ has a solution $x = a^b$, $a > 0$, $a \neq 1$, $x > 0$. The inequality $\log_a x < b$ when $a > 1$ has solutions $0 < x < a^b$, and when $0 < a < 1$, its solutions are $x > a^b$. The inequality $\log_a x > b$ when $a > 1$ is satisfied by $x > a^b$, and when $0 < a < 1$, its solutions are $0 < x < a^b$.

Example 1.1. Solve the equation

$$3^{3-x} - 27.$$

Proof.

$$3 - x = \log_3 27 = \log_3 3^3 = 3, \quad x = 0. \qquad \square$$

Example 1.2. Solve the inequality

$$\left(\sqrt{2} - 1\right)^x < \sqrt{2} + 1.$$

Proof. Note that $\left(\sqrt{2} - 1\right)\left(\sqrt{2} + 1\right) = 1$, i.e., $\left(\sqrt{2} + 1\right) = \left(\sqrt{2} - 1\right)^{-1}$ and $\sqrt{2} - 1 < 1$. Therefore,

$$x > \log_{\sqrt{2}-1}\left(\sqrt{2} + 1\right) \Longleftrightarrow x > -1. \qquad \square$$

Example 1.3. Solve the equation

$$\log_{\sqrt{3}}(x + 2) = 3.$$

Proof.

$$x + 2 = \sqrt{3}^3 = 3\sqrt{3} \Longleftrightarrow x = 3\sqrt{3} - 2. \qquad \square$$

1.3 Methods of Solving Exponential Equations and Inequalities

1.3.1 *Logarithm of both parts of an exponential equation or inequality*

Example 1.4. Solve the equation

$$3^x 2^{\frac{3x}{x+1}} = 36.$$

Proof. The area of definition of the functions included in the equation is $x \neq -1$. Taking logarithms from both sides of the equation, we get

$$x\log_2 3 + \frac{3x}{x+1} = 2 + 2\log_2 3,$$

$$\frac{x^2\log_2 3 + x\log_2 3 + 3x - 2x - 2 - 2x\log_2 3 - 2\log_2 3}{x+1} = 0,$$

$$\frac{x^2\log_2 3 - x(\log_2 3 - 1) - 2 - 2\log_2 3}{x+1} = 0.$$

The discriminant of the quadratic equation

$$x^2\log_2 3 - x(\log_2 3 - 1) - 2 - 2\log_2 3 = 0$$

is equal to $(\log_2 3 - 1)^2 + 8\log_2 3 + 8(\log_2 3)^2 = (3\log_2 3 + 1)^2$; therefore, it has roots

$$\left[\begin{array}{l} x_1 = \dfrac{\log_2 3 - 1 + 3\log_2 3 + 1}{2\log_2 3} = 2, \\[3mm] x_2 = \dfrac{\log_2 3 - 1 - 3\log_2 3 - 1}{2\log_2 3} = -1 - \dfrac{1}{\log_2 3} = -1 - \log_3 2. \end{array}\right.$$

Both of these numbers are different from -1 and are the roots of the original equation. $\square$

Example 1.5. Solve the inequality

$$x^{\lg x} > \frac{x^3}{100}.$$

Proof. The inequality is defined when $x > 0$. Taking the logarithms of base 10 from both parts, we get

$$\lg\left(x^{\lg x}\right) > \lg\left(\frac{x^3}{100}\right).$$

Using properties 3 and 5 of the logarithmic function, we reduce this inequality to the inequality

$$(\lg x)^2 > 3\lg x - \lg 100,$$

$$(\lg x)^2 - 3\lg x + 2 > 0,$$

hence $\lg x < 1$ or $\lg x > 2$. As a result, we get $0 < x < 10$ and $x > 100$. $\square$

1.3.2   *Substitution of variables in exponential equations and inequalities*

Example 1.6. Solve the inequality

$$2 \cdot 3^{x+1} - 6 \cdot 3^{x-1} - 3^x > 9.$$

Proof. Using properties 1 and 2 of the exponential function, we obtain

$$6 \cdot 3^x - 2 \cdot 3^x - 3^x > 9 \Longleftrightarrow 3^x > 3 \Longleftrightarrow x > 1. \qquad \square$$

Example 1.7. Solve the inequality

$$5^{7^x} < 7^{5^x}.$$

Proof. Taking the base 5 logarithms from both parts, we get the inequality

$$7^x < 5^x \log_5 7.$$

Dividing both parts of the inequality by a positive value 5^x, we get

$$\left(\frac{7}{5}\right)^x < \log_5 7$$

and since $\frac{7}{5} > 1$, $\log_5 7 > 0$,

$$x < \log_{\frac{7}{5}} \log_5 7. \qquad \square$$

Example 1.8. Solve the equation

$$49^x + 7^x - 2 = 0.$$

Proof. We denote $7^x = t$. Then, $49^x = (7^2)^x = 7^{2x} = (7^x)^2 = t^2$, according to property 5 of the exponential function, and we get the equation

$$t^2 + t - 2 = 0.$$

It has solutions $t = 1$, $t = -2$. Returning to the variable x,

$$7^x = 1 \quad \text{and} \quad 7^x = -2.$$

The first equation has a root $x = 0$, whereas the second one has no solutions. $\qquad \square$

Example 1.9. Solve the inequality

$$2^{x+2} - 2^{-x+2} < 15.$$

Proof. We denote $2^x = t$. Then, $2^{x+2} = 4t$ and $2^{-x+2} = \frac{4}{t}$. The inequality takes the form

$$4t + \frac{4}{t} < 15 \Longleftrightarrow \frac{4t^2 - 15t - 4}{t} < 0 \Longleftrightarrow 4t^2 - 15t - 4 < 0$$

since $t > 0$. Solving the inequality, we get $-\frac{1}{4} < t < 4$. Returning to the variable x,

$$-\frac{1}{4} < 2^x < 4.$$

The solution to these inequalities is $x < 2$. □

Example 1.10. Solve the homogeneous equation

$$3 \cdot 4^x + 2 \cdot 9^x = 5 \cdot 6^x.$$

Proof. Note that $4^x = (2^x)^2$, $9^x = (3^x)^2$, and $6^x = 2^x \cdot 3^x$. Dividing both parts of this equation by 9^x, we get the equivalent equation

$$3 \left(\frac{2}{3} \right)^{2x} - 5 \left(\frac{2}{3} \right)^x + 2 = 0.$$

Solving it, we get the equations

$$\left(\frac{2}{3} \right)^x = 1 \quad \text{and} \quad \left(\frac{2}{3} \right)^x = \frac{2}{3},$$

so $x = 0$ and $x = 1$. □

Let us consider a more complex example.

Example 1.11. For each value of a, solve the equation

$$2^{\frac{ax+3}{x^2+3}} + 2^{\frac{4x^2-ax+9}{x^2+3}} = 10.$$

Proof. The main idea of this problem is that the exponents in this equation satisfy the relation

$$\frac{ax+3}{x^2+3} + \frac{4x^2-ax+9}{x^2+3} = 4.$$

Therefore, if we denote

$$t = \frac{ax+3}{x^2+3}, \quad \text{then} \quad 4-t = \frac{4x^2-ax+9}{x^2+3}.$$

The equation will take the form

$$2^t + 2^{4-t} = 10 \iff 2^t + \frac{16}{2^t} = 10 \iff 4^t - 10 \cdot 2^t + 16 = 0,$$

so either $2^t = 2$ or $2^t = 8$. Therefore, $t = 1, t = 3$. Returning to the variable x, we get the equations

$$\frac{ax+3}{x^2+3} = 1 \quad \text{and} \quad \frac{ax+3}{x^2+3} = 3.$$

The first of these equations is equivalent to the equation $x^2 - ax = 0$, and for all values of a, it has the roots $x = 0$ and $x = a$. The second of these equations is equivalent to the equation $ax + 3 = 3(x^2 + 3)$, $3x^2 - ax + 6 = 0$. The discriminant of this quadratic equation is $a^2 - 72$; therefore, it has no solutions if $-6\sqrt{2} < a < 6\sqrt{2}$. If $a \le -6\sqrt{2}$ or $a \ge 6\sqrt{2}$, it has the roots

$$x_{1,2} = \frac{a \pm \sqrt{a^2-72}}{6}.$$

Combining the parts of the answer, we get that when $a \le -6\sqrt{2}$ or $a \ge 6\sqrt{2}$, the equation has the roots $x = 0, x = a$, $x_{1,2} = \frac{a\pm\sqrt{a^2-72}}{6}$, and when $-6\sqrt{2} < a < 6\sqrt{2}$, it has the roots $x = 0, x = a$.   □

Example 1.12. For what values of b does the equation

$$25^x - (2b+5)5^{x-1/x} + 10b \cdot 5^{-2/x} = 0$$

have two different roots?

Proof. The equation resembles the equation from Example 1.10. Indeed, when $x \neq 0$,

$$25^x = (5^x)^2, \quad 5^{-\frac{2}{x}} = \left(5^{-\frac{1}{x}}\right)^2, \quad 5^{x-\frac{1}{x}} = 5^x 5^{-\frac{1}{x}}.$$

Dividing both its parts by $5^{-\frac{2}{x}}$, we obtain an equivalent equation:

$$\left(5^{x+\frac{1}{x}}\right)^2 - (2b + 5)\, 5^{x+\frac{1}{x}} + 10b = 0.$$

Denote $5^{x+\frac{1}{x}} = t$, and consider the quadratic equation

$$t^2 - (2b + 5)\, t + 10b = 0.$$

The discriminant of this quadratic equation is $(2b + 5)^2 - 40b = (2b - 5)^2$. The roots of the equation are equal to $t = 2b, t = 5$. Returning to the variable x, we get the equations

$$\left[\begin{array}{l} 5^{x+\frac{1}{x}} = 2b, \\ 5^{x+\frac{1}{x}} = 5. \end{array} \right.$$

The second of these equations is equivalent to the equation $x + \frac{1}{x} = 1$. This equation has no roots. The first equation has no solutions when $b \le 0$. When $b > 0$, it takes the form

$$x + \frac{1}{x} = \log_5 2b \iff x^2 - (\log_5 2b) \cdot x + 1 = 0.$$

This equation has two roots if its discriminant $(\log_5 2b)^2 - 4 > 0$, or, in other words, if $\log_5 2b < -2$ or if $\log_5 2b > 2$, that is, when $0 < b < \frac{1}{50}$ or $b > 12\frac{1}{2}$. $\qquad\square$

Let's consider another equation that is essentially similar to the previous one.

Example 1.13. For what values of a does the equation

$$\left(\sqrt{x^2 - 3ax + 8} + \sqrt{x^2 - 3ax + 6}\right)^x$$
$$+ \left(\sqrt{x^2 - 3ax + 8} - \sqrt{x^2 - 3ax + 6}\right)^x = 2\left(\sqrt{2}\right)^x$$

have a single solution?

Proof. The domain of definition of this equation is given by the inequality $x^2 - 3ax + 6 \geq 0$. Note that

$$\left(\sqrt{x^2 - 3ax + 8} + \sqrt{x^2 - 3ax + 6}\right)^x \cdot \left(\sqrt{x^2 - 3ax + 8} - \sqrt{x^2 - 3ax + 6}\right)^x$$

$$= \left(x^2 - 3ax + 8 - (x^2 - 3ax + 6)\right)^x = 2^x.$$

Thus,

$$\left(\sqrt{x^2 - 3ax + 8} - \sqrt{x^2 - 3ax + 6}\right)^x = \frac{2^x}{\left(\sqrt{x^2 - 3ax + 8} + \sqrt{x^2 - 3ax + 6}\right)^x}.$$

The equation takes the form

$$\left(\sqrt{x^2 - 3ax + 8} + \sqrt{x^2 - 3ax + 6}\right)^x$$

$$+ \frac{2^x}{\left(\sqrt{x^2 - 3ax + 8} + \sqrt{x^2 - 3ax + 6}\right)^x} = 2\left(\sqrt{2}\right)^x,$$

or

$$\left(\sqrt{x^2 - 3ax + 8} + \sqrt{x^2 - 3ax + 6}\right)^{2x}$$

$$- 2\left(\sqrt{2}\right)^x \left(\sqrt{x^2 - 3ax + 8} + \sqrt{x^2 - 3ax + 6}\right)^x + \left(\sqrt{2}\right)^{2x} = 0.$$

This equation is reduced to the form

$$\left(\left(\sqrt{x^2 - 3ax + 8} + \sqrt{x^2 - 3ax + 6}\right)^x - \left(\sqrt{2}\right)^x\right)^2 = 0,$$

or

$$\left(\sqrt{x^2 - 3ax + 8} + \sqrt{x^2 - 3ax + 6}\right)^x - \left(\sqrt{2}\right)^x = 0.$$

The number $x = 0$ is obviously the root of this equation. If $x \neq 0$, then

$$\sqrt{x^2 - 3ax + 8} + \sqrt{x^2 - 3ax + 6} = \sqrt{2}.$$

Denote $x^2 - 3ax + 6 = t$. Then, the equation will take the form

$$\sqrt{t + 2} + \sqrt{t} = \sqrt{2}.$$

On the left side of this equation is an increasing function of the variable t, so it has a single solution, which is obviously $t = 0$, or

$$x^2 - 3ax + 6 = 0.$$

In order for the original equation to have a single solution $x = 0$, the equation $x^2 - 3ax + 6 = 0$ must have no solutions. This is equivalent to the condition $9a^2 - 24 < 0$, which gives $-\sqrt{\frac{8}{3}} < a < \sqrt{\frac{8}{3}}$. $\square$

Example 1.14. Solve the equation

$$\left(26 + 15\sqrt{3}\right)^x - 5\left(7 + 4\sqrt{3}\right)^x + 6\left(2 + \sqrt{3}\right)^x + \left(2 - \sqrt{3}\right)^x = 5.$$

Proof. A hint for the approach to the solution is contained in the equality

$$\left(2 + \sqrt{3}\right)^x \left(2 - \sqrt{3}\right)^x = (4 - 3)^x = 1.$$

It is important to note that $\left(2 + \sqrt{3}\right)^2 = 7 + 4\sqrt{3}$ and $\left(2 + \sqrt{3}\right)^3 = 26 + 15\sqrt{3}$.

Therefore, denoting $\left(2 + \sqrt{3}\right)^x = t$, we obtain the equation

$$t^3 - 5t^2 + 6t + \frac{1}{t} = 5.$$

Since $t > 0$, it is equivalent to the equation

$$t^2 - 5t + 6 - 5\frac{1}{t} + \frac{1}{t^2} = 0.$$

By replacing the variable $z = t + \frac{1}{t}$ in this equation and noting that

$$z^2 = t^2 + 2t \cdot \frac{1}{t} + \frac{1}{t^2} = t^2 + \frac{1}{t^2} + 2,$$

we get the equation

$$z^2 - 5z + 4 = 0.$$

It has the roots $z = 1$ and $z = 4$. Returning to the variable t,

$$\left[\begin{array}{l} t + \dfrac{1}{t} = 1, \\[2mm] t + \dfrac{1}{t} = 4. \end{array} \right.$$

The first of these equations has no solutions. The second reduces to the quadratic equation $t^2 - 4t + 1 = 0$, which has the roots $t = 2 + \sqrt{3}$ and $t = 2 - \sqrt{3}$. Returning to the variable x,

$$\left(2 + \sqrt{3}\right)^x = 2 + \sqrt{3} \Longleftrightarrow x = 1,$$

and

$$\left(2 + \sqrt{3}\right)^x = 2 - \sqrt{3} = \left(2 + \sqrt{3}\right)^{-1} \Longleftrightarrow x = -1. \qquad \square$$

Example 1.15. Solve the equation

$$x^{\log_7 4} - 5 \cdot 2^{\log_7 x} + 4 = 0.$$

Proof. We use the property 9 of logarithms:

$$x^{\log_7 4} = 4^{\log_7 x} = 2^{2\log_7 x}.$$

We denote $2^{\log_7 x} = z$ and get the equation

$$z^2 - 5z + 4 = 0.$$

It has the roots $z = 1$ and $z = 4$. Returning to the variable x,

$$\left[\begin{array}{l} 2^{\log_7 x} = 1, \\ 2^{\log_7 x} = 4, \end{array} \right.$$

so $\log_7 x = 0$ and $\log_7 x = 2$. It means that $x = 1$ and $x = 49$. $\square$

1.3.3   Standard logarithmic equations and inequalities

Example 1.16. Solve the equation

$$\log_3(x+1) + \log_3(x+3) = 1.$$

Proof. The domain of the equation definition is given by the inequality $x + 1 > 0$. Using equality 2 for logarithms, we obtain the equation

$$\log_3(x+1)(x+3) = 1, \quad x^2 + 4x + 3 = 3, \quad x = -4, \quad x = 0.$$

The condition $x + 1 > 0$ is satisfied only by $x = 0$. $\square$

Example 1.17. Solve the inequality

$$\log_{256} x^2 + \log_4 x + \log_2 x > 14.$$

Proof. The inequality is defined if $x > 0$. Using the equalities 4 and 5 for logarithms, we get

$$\log_{256} x^2 = \frac{1}{4} \log_2 x \quad \text{and} \quad \log_4 x = \frac{1}{2} \log_2 x.$$

Therefore,

$$\frac{7}{4} \log_2 x > 14 \iff \log_2 x > 8 \iff x > 256.$$ $\square$

Example 1.18. Solve the equation

$$\log_4 x + \log_3 x + \log_2 x = 1.$$

Proof. The equality is defined if $x > 0$. Using the equalities 5 and 6 for logarithms, we get

$$\log_4 x = \frac{1}{2}\log_2 x, \quad \log_3 x = \frac{\log_2 x}{\log_2 3},$$

from where, the original equation is equivalent to

$$\left(\frac{3}{2} + \frac{1}{\log_2 3}\right)\log_2 x = 1 \iff \left(\frac{3}{2} + \log_3 2\right)\log_2 x = 1 \iff \log_2 x = \frac{1}{\frac{3}{2} + \log_3 2},$$

$$x = 2^{\frac{1}{\frac{3}{2} + \log_3 2}}. \qquad \square$$

Example 1.19. Solve the inequality

$$\log_{x+14}\left(x^2 + 9x + 8\right) > 2.$$

Proof. We present a general scheme for solving inequalities of the form

$$\log_{a(x)} b(x) > k.$$

The inequality is defined under the conditions

$$a(x) > 0, \quad a(x) \neq 1, \quad b(x) > 0.$$

If $0 < a(x) < 1$, then we get $b(x) < (a(x))^k$, and if $a(x) > 1$, then $b(x) > (a(x))^k$.

Both of these cases are united by the inequality

$$(a(x) - 1)(b(x) - (a(x))^k) > 0,$$

which is equivalent to the original in the domain of definition.

Here, we have $a(x) = x + 14$, $b(x) = x^2 + 9x + 8$, $k = 2$. The domain of definition is set by the conditions

$$\begin{cases} x + 14 > 0, \\ x + 14 \neq 1, \\ x^2 + 9x + 8 > 0. \end{cases} \iff \left[\begin{array}{l} -14 < x < -13, \\ -13 < x < -8, \\ x > -1. \end{array}\right.$$

The inequality mentioned above takes the form

$$(x + 14 - 1)(x^2 + 9x + 8 - (x + 14)^2) > 0$$

$$\Longleftrightarrow (x + 13)(19x + 188) < 0 \Longleftrightarrow -13 < x < -\frac{188}{19}. \qquad \square$$

Example 1.20. Find all the values of a for which the inequality

$$\log_a \left(x^2 + 2\right) > 1$$

is true for all x.

Proof. The inequality is defined for all x and $a > 0$, $a \neq 1$. When $0 < a < 1$, it is equivalent to the inequality $x^2 + 2 < a$, which has no solutions. When $a > 1$, it is equivalent to the inequality $x^2 + 2 > a$, $x^2 > a - 2$, and the last inequality holds for all x if $a < 2$.

Answer: $1 < a < 2$. $\qquad \square$

Example 1.21. Find all the values of a for which the inequality

$$a \log_3 x + \log_{1/2} x > 1$$

has solutions, and there are no values large than 1 among the solutions.

Proof. Rewriting the inequality, we get $\log_3 x \cdot (a - \log_2 3) > 1$.

• If $a = \log_2 3$, then there are no solutions.
• If $a > \log_2 3$, then $x > 3^{\frac{1}{a - \log_2 3}} > 1$.
• If $a < \log_2 3$, then $0 < x < 3^{\frac{1}{a - \log_2 3}} < 1$.

Finally, we get $a < \log_2 3$. $\qquad \square$

1.3.4 *Finding a multiplicative dependency*

Example 1.22. Solve the equation

$$\log_{\sqrt{2} + \sqrt{5} + \sqrt{7}} (x) + \log_x \left(\sqrt{2} + \sqrt{5} - \sqrt{7}\right) = \frac{3}{2} + \log_x 2\sqrt{10}.$$

Proof. The domain of definition of the equation is $x > 0$, $x \neq 1$. The idea of the solution is the equality

$$\left(\sqrt{2} + \sqrt{5} + \sqrt{7}\right)\left(\sqrt{2} + \sqrt{5} - \sqrt{7}\right) = \left(\sqrt{2} + \sqrt{5}\right)^2 - 7 = 2\sqrt{10}.$$

Therefore, the original equation can be rewritten as

$$\log_{\sqrt{2}+\sqrt{5}+\sqrt{7}} x + \log_x \left(\sqrt{2} + \sqrt{5} - \sqrt{7}\right)$$
$$= \frac{3}{2} + \log_x \left(\sqrt{2} + \sqrt{5} + \sqrt{7}\right)\left(\sqrt{2} + \sqrt{5} - \sqrt{7}\right),$$

or

$$\log_{\sqrt{2}+\sqrt{5}+\sqrt{7}} x = \frac{3}{2} + \log_x \left(\sqrt{2} + \sqrt{5} + \sqrt{7}\right).$$

Denote

$$t = \log_{\sqrt{2}+\sqrt{5}+\sqrt{7}} x.$$

Then, we have

$$\log_x \left(\sqrt{2} + \sqrt{5} + \sqrt{7}\right) = \frac{1}{t},$$

and the equation in question will take the form $t - \frac{3}{2} - \frac{1}{t} = 0$, so we find the roots $t = 2$, $t = \frac{1}{2}$.
Returning to the variable x, we get

$$x = \left(\sqrt{2} + \sqrt{5} + \sqrt{7}\right)^2, \quad x = \left(\sqrt{2} + \sqrt{5} + \sqrt{7}\right)^{-\frac{1}{2}}. \qquad \square$$

Example 1.23. Solve the system of equations

$$\begin{cases} 6x^2 + 17xy + 7y^2 = 16, \\ \log_{2x+y} (3x + 7y) = 3. \end{cases}$$

Proof. The domain of definition of the system of equations is defined by the inequalities $3x + 7y > 0$, $2x + y > 0$, and $2x + y \neq 1$. Here, the idea of the solution is to use the identity

$$6x^2 + 17xy + 7y^2 = (3x + 7y)(2x + y)$$

and the equality, following from the second equation,

$$3x + 7y = (2x + y)^3.$$

As a result, we get that $(2x + y)^4 = 16$ and $2x + y = 2$ since $2x + y > 0$; therefore, the original system of equations is equivalent to the system

$$\begin{cases} 2x + y = 2, \\ 3x + 7y = 8, \end{cases}$$

and we get $x = \frac{6}{11}$, $y = \frac{10}{11}$.   □

Example 1.24. Solve the equation

$$\log_{3x+7}\left(4x^2 + 12x + 9\right) + \log_{2x+3}\left(6x^2 + 23x + 21\right) = 4.$$

Proof. The equation is defined under the conditions

$$\begin{cases} 3x + 7 > 0, \\ 2x + 3 > 0, \\ 3x + 7 \neq 1, \\ 2x + 3 \neq 1. \end{cases} \iff \begin{cases} x > -\dfrac{3}{2}, \\ x \neq -1. \end{cases}$$

Since $4x^2 + 12x + 9 = (2x + 3)^2$ and $6x^2 + 23x + 21 = (2x + 3)(3x + 7)$, then the same equalities allow us to bring the original equation to the form

$$2\log_{3x+7}(2x + 3) + 1 + \log_{2x+3}(3x + 7) = 4.$$

We use the equality that is true in the domain of definition

$$\log_{3x+7}(2x + 3) = \frac{1}{\log_{2x+3}(3x + 7)},$$

and denote $t = \log_{2x+3}(3x + 7)$. We get the equation

$$\frac{2}{t} + t = 3,$$

solving which we get the roots $t_1 = 1$ and $t_2 = 2$. Returning to the variable x, we get one of the equations as

$$\log_{2x+3}(3x+7) = 1 \iff 3x+7 = 2x+3 \iff x = -4.$$

This root is not in the domain of the equation definition. The second equation gives

$$\log_{2x+3}(3x+7) = 2 \iff 3x+7 = (2x+3)^2 \iff x = -2, \quad x = -\frac{1}{4}.$$

The domain of definition includes only $x = -\frac{1}{4}$. $\qquad\square$

1.3.5 Using the properties of logarithmic and exponential functions

Example 1.25. Solve the inequality

$$\log_2(x+2) > 1 - x.$$

Proof. The function $\log_2(x+2)$ increases in the interval $(-2, +\infty)$, and the function $1 - x$ decreases in the interval $(-\infty, +\infty)$. For $x = 0$, both of these functions are equal to 1. Therefore, the inequality holds for $x > 0$. $\quad\square$

Example 1.26. The inequality equivalent to the one considered is

$$2^x > 1 - x.$$

Proof. The function 2^x increases in the interval $(-\infty, +\infty)$, and the function $1 - x$ decreases in the interval $(-\infty, +\infty)$. For $x = 0$, both of these functions are equal to 1. Therefore, the inequality holds for $x > 0$. $\quad\square$

Example 1.27. Solve the inequality

$$(\log_2|2x|)^2 - 5\log_2|2x| + |2x|\log_2|2x| - 4|x| + 6 \geq 0.$$

Proof. The inequality is defined at $x \neq 0$. Consider it as a quadratic inequality with respect to $\log_2|2x|$. The coefficients depend on x, and the discriminant is

$$(|2x| - 5)^2 - 4(-4|x| + 6) = 4x^2 - 4|x| + 1 = (|2x| - 1)^2.$$

The roots of the corresponding equation are 2 and $-|2x|+3$, and the original inequality can be rewritten as

$$(\log_2 |2x| - 2)(\log_2 |2x| + 2|x| - 3) \geq 0.$$

The first parenthesis is positive if $|x| > 4$, equal to 0 if $|x| = 4$, and negative if $0 < |x| < 4$. The sign of the second parenthesis is easy to determine by analogy with the previous example. The function inside the parenthesis is an increasing function of the variable $|x|$ and vanishes if $|x| = 1$. Therefore, the second parenthesis is positive if $|x| > 1$, equal to 0 if $|x| = 1$, and negative if $0 < |x| < 1$. So, the product of the parentheses is non-negative if $0 < |x| \leq 1$ or $|x| \geq 4$. Thus, we get the answer: $(-\infty, -4] \cup [-1, 0) \cup (0, 1] \cup [4, +\infty)$.   □

The following example is devoted to solving a difficult problem.

Example 1.28. Solve the equation

$$\log_2 (4x + 1) \log_5 (4x + 4) + \log_3 (4x + 2) \log_4 (4x + 3)$$
$$= 2 \log_3 (4x + 2) \log_5 (4x + 4).$$

Proof. Denote $z = 4x - 1$, and rewrite the equation as

$$\log_2 (z + 2) \log_5 (z + 5) + \log_3 (z + 3) \log_4 (z + 4)$$
$$= 2\log_3 (z + 3) \log_5 (z + 5).$$

Its domain of definition is $z > -2$. In this case, $\log_3 (z + 3) \log_5 (z + 5) \neq 0$, and convert the original equation to the equivalent

$$\frac{\log_2 (z + 2)}{\log_3 (z + 3)} + \frac{\log_4 (z + 4)}{\log_5 (z + 5)} = 2.$$

Denote by $f(z)$ the left part of the last equation. See in Fig. 1.5 graph of this function. It is clear that $z = 0$ is the solution to this equation. We prove that for $z > 0$, the left side of this equation is greater than 2, and for $0 > z > -2$, it is less than 2. To do this, we prove that for any $k > 1$, the inequalities are satisfied:

-
$$\frac{\log_k (z + k)}{\log_{k+1} (z + k + 1)} > 1,$$

if $z > 0$,

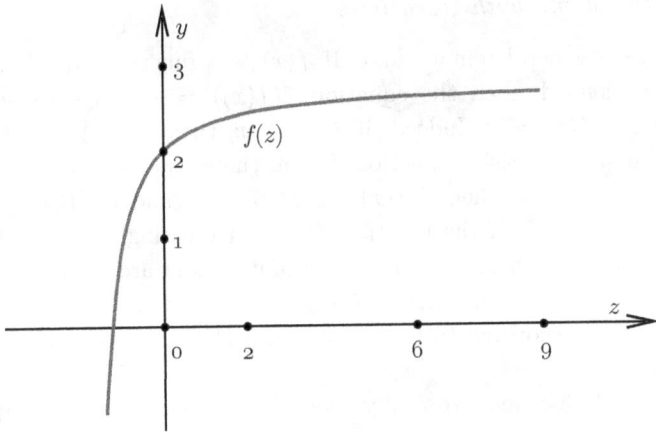

Fig. 1.5. Graph of $f(Z)$.

- $$\frac{\log_k(-z+k)}{\log_{k+1}(-z+k+1)} < 1,$$

if $k > z > 0$.

The first inequality to be proved for $z > 0$ is equivalent to the inequality

$$\log_k(z+k) > \log_{k+1}(z+k+1),$$

which is equivalent to

$$1 + \log_k\left(\frac{z}{k}+1\right) > 1 + \log_{k+1}\left(\frac{z}{k+1}+1\right),$$

or

$$\log_k\left(\frac{z}{k}+1\right) > \log_k\left(\frac{z}{k+1}+1\right) > \log_{k+1}\left(\frac{z}{k+1}+1\right).$$

Quite similarly, the second inequality being proved is equivalent to the inequality

$$\log_k(-z+k) < \log_{k+1}(-z+k+1)$$

and follows from the inequalities

$$\log_k\left(-\frac{z}{k}+1\right) < \log_k\left(-\frac{z}{k+1}+1\right) < \log_{k+1}\left(-\frac{z}{k+1}+1\right).$$

Thus, we have proved that $z = 0$ is the only solution to this equation. Therefore, $x = \frac{1}{4}$. □

1.3.6 *Problems with iterations*

Let's make a general remark first. If $f(x)$ is a function and if the value $f(f(x))$ is defined, then the equation $f(f(x)) = x$ is a consequence of the equation $f(x) = x$. Indeed, if $f(x) = x$, then $f(f(x)) = f(x) = x$. Let $f(x)$ be an increasing function. Then, these equations are equivalent. Indeed, if $f(x) < x$, then $f(f(x)) < f(x) < x$, and if $f(x) > x$, then $f(f(x)) > f(x) > x$. If the function $f(x)$ is decreasing, then the function $f(f(\cdots f(x))\cdots)$, where an odd number of iterations are carried out, and it also decreases, so the equation $f(f(\cdots f(x))\cdots) = x$ has a unique solution that coincides, according to the above, with the solution of the equation $f(x) = x$.

Note that for monotonically decreasing functions, the equations $f(x) = x$ and $f(f(x)) = x$ may no longer be equivalent. We can consider as an example the function $f(x) = -x^3$. Then, the equation $f(x) = x$ has a single solution, $x = 0$, and the equation $f(f(x)) = x$ or $x^9 = x$ has three solutions: $x = -1$, 0, and 1.

Example 1.29. Solve the equation

$$1 - \log_{10}\left(1 - \log_{10}((1 - \log_{10} x))\right) = x.$$

Proof. The function $1 - \log_{10}x$ is defined for $x > 0$ and is decreasing. Therefore, the equation in question is equivalent to the equation

$$1 - \log_{10}x = x.$$

The solution of this equation is almost analogous to the solution of Example 24; the answer is $x = 1$.   □

Example 1.30. Solve the equation

$$2^{1-2^{1-2^x}} = 1 - x.$$

Proof. Note first that if x is the solution of the equation

$$2^x = 1 - x, \quad x = 1 - 2^x,$$

then

$$2^{1-2^{1-2^x}} = 2^{1-2^x} = 2^x = 1 - x.$$

The function $1 - 2^x$ decreases. Therefore, the solutions of the original equation coincide with the solutions of equation $2^x = 1 - x$, from whence $x = 0$. $\square$

Example 1.31. Solve the equation

$$1 + \ln\left(1 + \ln x\right) = x.$$

Proof. The function $1 + \ln x$ is defined for $x > 0$ and is increasing. Therefore, the equation in question is equivalent to the equation

$$1 + \ln x = x.$$

Since the tangent for the function $\ln x$ at point 1 will be $x - 1$ and the function $\ln x$ is concave down, then the solution $x = 1$ is unique. $\square$

Example 1.32. Solve the equation

$$2^{2^x - 1} = x + 1.$$

Proof. The function $f(x) = 2^x - 1$ increases. Since the original equation is equivalent to

$$2^{2^x - 1} - 1 = x \iff f(f(x)) = x,$$

its solutions coincide with the solutions of the equation $f(x) = x$. Let's solve the equation $2^x = x + 1$. Since the function $y = 2^x$ is a convex function, the intersection with the line $y = x + 1$ has at most two solutions. However, it is easy to see that $x = 0$ and $x = 1$ are solutions to the equation $2^x = x + 1$. Since we have found two solutions, there are no more solutions. $\square$

Problems

1.1. Solve the inequality $\dfrac{1}{2} \cdot \log_3 x^2 \geq \dfrac{1}{3} \cdot \log_3(-x^3)$.

1.2. Solve the inequality $\dfrac{(\log_3 5)^x - (\log_3 5)^3}{(\log_3 5)^{-x} - x(\log_{3^x} 5)} > 0$.

1.3. Solve the inequality $\log_{\frac{x-2}{2x-10}}\left(\dfrac{x+2}{4}\right) \leq 1$.

1.4. Solve the inequality

$$\sqrt{\log_5 x + 3} - \sqrt{\log_5 x - 2} < \sqrt{\log_5 x - 1}.$$

1.5. Solve the inequality $\log_{x+4}(x^2 + 9x + 8) > 2$.

1.6. Solve the inequality

$$2\log_\pi(\sin x)\log_\pi(\sin 2x) - \log_\pi^2(\sin 2x) \leq \log_\pi^2(\sin x).$$

1.7. Solve the inequality $\log_{\frac{3-x}{2}}\left(\dfrac{6}{x+1}\right) \geq -1$.

1.8. Solve the inequality

$$\log_{-2-x}(-3 - 2x) \geq \log_{-2-x}(-3x/2).$$

1.9. Solve the inequality

$$\log_{\sqrt{2}}(6 - x - x^2) + \log_2(x^2 - 2x + 1) + 2$$
$$> 2\log_4(x^2 - 4x + 3)^2.$$

1.10. Solve the inequality $\log_{4x^2} x^2 \cdot \log_{8x^4} x^4 \leq 1$.

1.11. Solve the inequality

$$x \geq \log_2(101 \cdot 10^x - 10^{2+2x}) - \log_5(101 \cdot 2^x - 5^{2+x} \cdot 2^{2+2x}).$$

1.12. Solve the inequality

$$(\log_{3+x}(1 - 2x))(\log_{1-2x} x^2) \leq (\log_{3+x}(1 - 3x))(\log_{1-3x}(2 - x)).$$

1.13. Solve the inequality

$$x^{\log_5 9} + 7 \cdot 3^{\log_5 x} - 11 = 0.$$

1.14. Solve the inequality

$$(3 - x)\log_3(1 + \sqrt{7})^{x^2+3x+2} > \sqrt{2 - x}\log_3(8 + 2\sqrt{7})^{(x+1)\sqrt{x+1}}.$$

1.15. Solve the inequality
$$\log_{\sqrt{2}+\sqrt{3}+\sqrt{5}} x + \log_x(\sqrt{2}+\sqrt{3}-\sqrt{5}) = 3/2 + \log_x(2\sqrt{6}).$$

1.16. Solve the inequality
$$\log_{8-7x}(x^3 - 3x^2 - 37x/8 + 55/8) + 2\log_{(8-7x)^2}(x+3) = 1.$$

1.17. Solve the inequality
$$(1/2) \cdot \log_{x-1}(x^2 - 8x + 16) + \log_{4-x}(-x^2 + 5x - 4) > 3.$$

1.18. Solve the inequality
$$\log_2(\sqrt{2x^2 - 1} + |x|) \le \log_{|x+3|}(x^2 - 1) \cdot \log_2 |x+3|.$$

1.19. Solve the inequality
$$\log_3^2 |x| - 9\log_3 |2x| + 2|x| \cdot \log_3 |x| - 4|x| + 14 \ge 0.$$

1.20. Solve the inequality
$$\log_{4-\frac{x^2}{2}}\left(55 - \frac{x^2}{2} + \frac{x}{2}\right) \le \frac{2}{\log_{5+\sqrt{3}} 2 + \log_{5+\sqrt{3}}(14+5\sqrt{3})}.$$

1.21. Solve the inequality
$$\log_4(16(x-2)^2) \cdot \log_{1/16}^2 \frac{(x-2)^4}{64}$$
$$- \frac{5}{4}\log_{64}(x^3 - 6x^2 + 12x - 8)^2 < \frac{15}{2}.$$

1.22. Solve the equation
$$\log_2(3x+1)\log_5(3x+4) + \log_3(3x+2)\log_4(3x+3)$$
$$= 2\log_3(3x+2)\log_5(3x+4).$$

1.23. Solve the equation
$$(26 + 15\sqrt{3})^x - 3(7 + 4\sqrt{3})^x - 2(2 + \sqrt{3})^x + (2 - \sqrt{3})^x = 3.$$

1.24. Find all values of b for which the inequality
$$\log_b(x^2 + 5) > 1$$
is executed for all values of x.

1.25. For each value of c, solve the equation

$$4^x + c \cdot 25^x = 3 \cdot 10^x.$$

1.26. Find all the values of a for each of which the inequality

$$a \log_5 x + \log_{1/3} x > 1$$

has solutions, and there are no values less than 1 among the solutions.

1.27. For any valid value of a, solve the inequality

$$\log_{2a}(\log_3 x^2) > 1,$$

and find at what value a the set of points x that are not a solution to the inequality represents a gap whose length is 6?

1.28. For which values of p is the ratio of the sum of the coefficients of the polynomial $(px^2 - 7)^{18}$ to its free member minimally?

1.29. For each valid value of b, solve the inequality

$$\sqrt{7 + \log_b x^2} + (\log_b |x|)(1 + 2\log_x b) > 0.$$

1.30. For each value of a, solve the inequality

$$\log_{1/9}(x^2 - 6x - a^2 - 5a + 12) < -1,$$

and find for what values of a the set of numbers that are not solutions to this inequality is a segment of the numerical axis whose length is less than $2\sqrt{3}$.

1.31. For each value of a, solve the equation

$$3^{\frac{ax+2}{x^2+2}} + 3^{\frac{3x^2-ax+4}{x^2+2}} = 12.$$

1.32. Find all the values of a for each of which the equation

$$\log_{a-6,5}(x^2 + 1) = \log_{a-6,5}((a-5)x)$$

has exactly two different solutions.

1.33. Find all the values of a for each of which the equation

$$\frac{(x^3 - 1)(x^2 - 16)}{\lg(15a - x) - \lg(x - a)} = 0$$

has a unique solution.

1.34. Find the values of p for which the equation

$$4(x - \sqrt{p \cdot 7^p})x + p + 7(7^p - 1) = 0$$

has roots, and what are the signs of the roots for different values of p?

1.35. For which values of b the equation

$$4^x - (a + 2)2^{x-1/x} + 2b \cdot 2^{-2/x} = 0$$

has exactly two different solutions.

1.36. Find all the values of a for each of which the equation

$$\log_{a+1} x + \log_x(19 - 8a) = 2$$

has at least two roots and, at the same time, the product of all its roots is at least 0.01.

1.37. For the three values of a, -1.2, -0.67, and -0.66, find all those values for each of which the equation

$$\left(2^{a+4} + 15(x + a)\right) \cdot \left(1 + 2\cos\left(\pi\left(a + \frac{x}{2}\right)\right)\right) = 0$$

has at least one solution satisfying the condition $0 \le x \le 1$.

1.38. Solve the equation

$$\sqrt[3]{2 - x} = 2 - (2 - x^3)^3.$$

1.39. Solve the equation

$$\log_2(1 + \log_2 x) = x - 1.$$

1.40. Let $f(x) = x/3 + 2$. Find the value of the function

$$\underbrace{f(f(\ldots f(x) \ldots))}_{2009-\text{times}}$$

at the point $x = 4$.

1.41. Solve the equation

$$f(f(x)) = f(x),$$

where

$$f(x) = 2^{-x^3 - x} - 5.$$

Hints and Answers

1.1. $x < 0$.

1.2. $(-1; 0) \cup (0; 3)$.

1.3. $x \in [6; 8)$.

1.4. $x > 5^{2\sqrt{21}/3}$.

1.5. $(8; +\infty)$.

1.6. $2\pi n < x < \pi/2 + 2\pi n,\ n \in \mathbb{Z}$.

1.7. $(-1; 1) \cup [2; 3)$.

1.8. $(-\infty; -6] \cup (-3; -2)$.

1.9. $((-1 - \sqrt{73})/4; 1) \cup (1; (-1 + \sqrt{73})/4)$.

1.10. $(-\infty; -2^{-3/7}] \cup (-2^{-3/4}; 1/2) \cup (1/2; 2^{-3/4}) \cup [2^{-3/7}; +\infty)$.

1.11. $(-\infty; -2] \cup [0; \lg 101 - 2)$.

1.12. $(-3; -2) \cup (-2; 0) \cup (0; 1/3)$.

1.13. $5^{\log_3(-7+\sqrt{93})/2}$.

1.14. $(-1; 2]$.

1.15. $(\sqrt{2} + \sqrt{3} + \sqrt{5})^2,\ (\sqrt{2} + \sqrt{3} + \sqrt{5})^{-1/2}$.

1.16. $x = -1$.

1.17. $(2; 5/2) \cup (5/2; 3)$.

1.18. $(-\infty; -4) \cup (-4; -3) \cup (-3; -1 - \sqrt{3}] \cup [1 + \sqrt{3}; +\infty)$.

1.19. $(-\infty; -9] \cup [-3; 0) \cup (0; 3] \cup [9; +\infty)$.

1.20. $x \in (-\sqrt{8}; -\sqrt{6}) \cup (\sqrt{6}; \sqrt{8})$.

1.21. $(-6; 3/2) \cup (3/2; 2) \cup (2; 5/2) \cup (5/2; 10)$.

1.22. $1/3$.

1.23. ± 1.

1.24. $b \in (1; 5)$.

1.25. $c \in [1; 5/2) \cup [4; +\infty)$. If $c \leq 0$, $x = \log_{2/5}((3 + \sqrt{9 - 4c})/2)$; if $0 < c \leq 9/4$, $x \log_{2/5}((3 \pm \sqrt{9 - 4c})/2)$; else if $c > 9/4$, there are no solutions.

1.26. $a > \log_3 5$.

1.27. (1) If $a \in (0; 1/2)$, then $x \in (-3^a; -1) \cup (1; 3^a)$; if $a > 1/2$, then $x \in (-\infty; -3^a) \cup (3^a; +\infty)$; and (2) $a = 1$.

1.28. $p = 7$. **Hint:** The sum of the coefficients of any polynomial is equal to its value at the point 1.

1.29. If $b \in (0; 1)$, then $x \in (0; 1) \cup (1; b^{-3})$; if $b \in (1; +\infty)$, then $x \in (b^{-3}; 1) \cup (1; +\infty)$.

1.30. (1) For $a \in (-2; -3)$, $x \in \mathbb{R}$, for $a \in (-\infty; -2] \cup [-3; +\infty)$, $x \in (-\infty; 3 - \sqrt{a^2 + 5a + 6}) \cup (3 + \sqrt{a^2 + 5a + 6}; +\infty)$; and (2) $a \in ((-5 - \sqrt{13})/2; -3) \cup (-2; (-5 + \sqrt{13})/2)$.

1.31. $x = 0$, for any a, $x = (a \pm \sqrt{a^2 - 16})/4$ if $|a| \geq 4$.

1.32. $a \in (7; 7.5) \cup (7.5; +\infty)$.

1.33. $a \in (1/15; 1/8) \cup (1/8; 4/15] \cup \{1/2\} \cup [1; 4)$.

1.34. If $p = 0$, then $x = 0$; if $p = 7$, then $x = 7^4/2$; else if $p > 7$, we have two positive roots.

1.35. $a \in (0; 1/4) \cup (4; +\infty)$.

1.36. $a \in [-9/10; 0) \cup (2; 9/4) \cup (9/4; 19/8)$.

1.37. $a = -1.2$; $a = -0.67$.

1.38. 1.

1.39. 1, 2.

1.40. $3 + 3^{-2009}$.

1.41. -1.

Chapter 2

Equations and Inequalities with Polynomials and Rational Functions

2.1 Basic Definitions and Statements

Definition 2.1. A *polynomial* is a function of x of the form

$$a_n x^n + \cdots + a_1 x + a_0,$$

where $a_n, \ldots, a_0$ are real numbers, i.e., $a_k \in \mathbb{R}$, $k = 0, 1, \ldots, n$. The number n is called *the degree of the polynomial*. The number x_0 is called the *root of the polynomial* if

$$a_n x_0^n + \cdots + a_1 x_0 + a_0 = 0.$$

If we consider only real numbers, then not every polynomial has a root, for example, the equation

$$x^2 + 1 = 0$$

has no real solutions.

If we allow the consideration of complex numbers, then any polynomial of degree n with complex coefficients has n roots (taking multiplicity into account). However, we further consider only polynomials with real coefficients and only their real roots.

The linear equation

$$ax + b = 0,$$

provided that $a \neq 0$, has a single root $x = -\frac{b}{a}$. If $a = 0$, then it either has no roots at $b \neq 0$ or its roots are all real numbers at $b = 0$.

Let us consider the quadratic equation

$$ax^2 + bx + c = 0. \tag{2.1}$$

We assume that $a \neq 0$; otherwise, it reduces to the linear equation already considered. Convert it to the form

$$a\left(x + \frac{b}{2a}\right)^2 = \frac{b^2 - 4ac}{a} \Longleftrightarrow \left(x + \frac{b}{2a}\right)^2 = \frac{b^2 - 4ac}{a^2},$$

and we get that if $b^2 - 4ac < 0$, then the equation has no solutions.

- If $b^2 - 4ac = 0$, then the solution is unique, $x = -\frac{b}{2a}$ (this root has a multiplicity of 2).
- If $b^2 - 4ac > 0$, then the roots of the original equation are the numbers

$$x_1 = \frac{-b + \sqrt{b^2 - 4ac}}{2a}, \quad x_2 = \frac{-b - \sqrt{b^2 - 4ac}}{2a}.$$

Definition 2.2. The value $b^2 - 4ac$ is called the *discriminant* of equation (2.1).

Remark 2.1.

1. An important role in solving problems of quadratic equations is played by **Vieta's theorem**. For the quadratic equation $ax^2 + bx + c = 0$, $a \neq 0$, where x_1, x_2 are roots of the equation (case $D \geq 0$), $ax^2 + bx + c = a(x - x_1)(x - x_2)$. This implies the following Vieta's formulas:

$$x_1 + x_2 = -\frac{b}{a};$$

$$x_1 x_2 = \frac{c}{a}.$$

2. The second important point is that when solving problems that reduce to the study of quadratic equations, one must remember the geometric interpretation of the quadratic equation. For example, selecting a full square, we get (case $a \neq 0$)

$$ax^2 + bx + c = a \cdot \left(x + \frac{b}{2a}\right)^2 + \left(c - \frac{b^2}{4a}\right)$$

$$= a \cdot (x - x_v)^2 + y_v,$$

where x_v and y_v are defined by

$$x_v = -\frac{b}{2a}, \quad y_v = c - \frac{b^2}{4a}.$$

Whence it follows that the point is the vertex of the parabola $(x_v; y_v)$. When $a > 0$, the branches of the parabola are directed upward, and the vertex point of the parabola is the minimum point. When $a < 0$, the branches of the parabola are directed downward, with the vertex point of the parabola being the maximum point.

2.2 Solutions in Rational Numbers of Equations with Integer Coefficients

Let

$$a_n x^n + a_{n-1} x^{n-1} + \cdots + a_1 x + a_0 = 0, \tag{2.2}$$

where $a_0, a_1, \ldots, a_{n-1}, a_n$ are integers, $a_n \neq 0$. We formulate a theorem that allows us to find those of its roots that are rational numbers.

Theorem 2.1. *Let the number* $\alpha = p/q$, *where* p, q *are integers,* $q > 0$, *and* p *and* q *are mutually prime (i.e., their greatest common divisor is 1), be the root of equation* (2.2). *Then,* p *divides the number* a_0, *and* q *divides* a_n.

Corollary 2.1. *If in equation* (2.2), $a_n = 1$, *then* $q = 1$, *and the integer* $\alpha = p$, *which is the root of such an equation, must divide* a_0.

Proof. To prove the theorem, we substitute $\alpha = p/q$ into equation (2.2) and get the equality

$$a_n \frac{p^n}{q^n} + a_{n-1} \frac{p^{n-1}}{q^{n-1}} + \cdots + a_1 \frac{p}{q} + a_0 = 0.$$

Multiplying both its parts by q^n, we get

$$a_n p^n + a_{n-1} p^{n-1} q + \cdots + a_1 p q^{n-1} + a_0 q^n = 0. \tag{2.3}$$

It follows from this equality that

$$a_n p^n + a_{n-1} p^{n-1} q + \cdots + a_1 p q^{n-1} = -a_0 q^n. \tag{2.4}$$

The left part of equality (2.4) is divisible by the number p. Therefore, the number $-a_0 q^n$ is also divisible by p, and since q has no common divisors with p, the number a_0 is divisible by p. Similarly, by rewriting equality (2.3) as

$$a_{n-1}p^{n-1}q + \cdots + a_1 p q^{n-1} + a_0 q^n = -a_n p^n,$$

we get that a_n is divisible by q.   □

Example 2.1. Solve the equation

$$x^3 - 6x^2 + 15x - 14 = 0. \qquad (2.5)$$

Proof. As a corollary to the theorem, all rational roots of equation (2.5) must be integers and divide the number 14. Thus, we look for the root among the numbers ± 1, ± 2, ± 7, and ± 14. Substituting these numbers into equation (2.5), we find that for $x = 2$ holds. To find the remaining roots, you can use the following technique. Let us put $x = t + 2$. Then, equation (2.5) becomes the equation

$$t^3 + 3t = 0, \quad \text{or} \quad t(t^2 + 3) = 0.$$

The root $t = 0$ corresponds to the root $x = 2$ of the original equation, and the equation

$$t^2 + 3 = 0$$

has no roots.   □

Example 2.2. Solve the equation

$$2x^3 + x^2 + x - 1 = 0. \qquad (2.6)$$

Proof. According to the theorem, the rational roots of $\alpha = p/q$ of equation (2.6) can only be the numbers $\pm \frac{1}{2}$ and ± 1. By direct substitution, we find that the number $x = \frac{1}{2}$ is indeed the root of this equation. Next, acting as in Example (2.1), we have

$$2x^3 + x^2 + x - 1 = (2x - 1)(x^2 + x + 1) = 0.$$

The equation $x^2 + x + 1 = 0$ has no roots.   □

Remark 2.2. The division of a polynomial by a monomial can be implemented using the so-called *Gorner scheme*. We describe this scheme.

The polynomial $P_n(x) = a_n x^n + a_{n-1} x^{n-1} + a_{n-2} x^{n-2} + \cdots + a_1 x + a_0$ must be divided by $x - a$, i.e., represented in the form of

$$P_n(x) = (x - a)Q_{n-1}(x) + R,$$

where $R = P_n(a)$ is the remainder and $Q_{n-1}(x) = b_{n-1} x^{n-1} + b_{n-2} x^{n-2} + \cdots + b_1 x + b_0$ is a polynomial. The coefficients b_k, $k = 0.1, \ldots, n - 1$ and the remainder R are conveniently calculated using the table

	a_n	a_{n-1}	a_{n-2}	$\cdots$		$\cdots$	a_2	a_1	a_0
a	b_{n-1}	b_{n-2}	b_{n-3}	$\cdots$		$\cdots$	b_1	b_0	R

where

$$b_{n-1} = a_n;$$

$$b_{n-2} = a_{n-1} + a \cdot b_{n-1};$$

$$b_{n-3} = a_{n-2} + a \cdot b_{n-2};$$

$$\cdots \; ;$$

$$b_0 = a_1 + a \cdot b_1;$$

$$R = a_0 + a \cdot b_0.$$

Example 2.3. Prove that the system of equations

$$\begin{cases} 8x^3 + 18x^2 + 15x + 14 = 0, \\ (10 + 4x)^y - 2 = y(5 + 7/x) + 7^{x+y} \cdot \sqrt{16x(x+1)^2 + 40x^2 + 89x + 49} \end{cases}$$

has at least two different solutions.

Proof. Let us examine the zeros of the equation $8x^3 + 18x^2 + 15x + 14 = 0$. All the divisors of a_3 are ± 1, ± 2, ± 4, and ± 8, and the divisors of a_0 are ± 1, ± 2, ± 7, and ± 14. According to the theorem on rational solutions of equations, we conclude that all rational zeros of the equation $8x^3 + 18x^2 + 15x + 14 = 0$ are among the following numbers:

$$\pm 1, \pm 2, \pm 7, \pm 14,$$

$$\pm 1/2, \pm 2/2, \pm 7/2, \pm 14/2,$$

$$\pm 1/4, \pm 2/4, \pm 7/4, \pm 14/4,$$

$$\pm 1/8, \pm 2/8, \pm 7/8, \pm 14/8.$$

But noting that the function $f(x) = 8x^3 + 18x^2 + 15x + 14$, for positive x, is strictly greater than zero, all rational solutions of the equation $f(x) = 0$ are among the following numbers:

$$-1/8, -1/4, -1/2, -1, -7, -7/2, -7/4, -7/8, -14.$$

The numbers of the form $-1/8, -7/8$ cannot be a solution of the equation $f(x) = 0$ because for these numbers x, $8 \cdot 4 \cdot (18x^2 + 15x + 14) \in \mathbb{Z}$, and $8 \cdot 4 \cdot (8x^3) \notin \mathbb{Z}$, so

$$8 \cdot 4 \cdot (8x^3 + 18x^2 + 15x + 14) \notin \mathbb{Z}$$

and, in particular, $f(x) \neq 0$.

So, all rational solutions of the equation $f(x) = 0$ are among the following numbers:

$$-1/4, -1/2, -1, -7, -7/2, -7/4, -14.$$

Let us start checking in turn whether these numbers are the roots of the equation $f(x) = 0$. Let us start with the number $x = -1/4$:

	8	18	15	14
-1/4	8	16	11	45/4

Since the remainder of the division is $45/4$, $x = -1/4$ is not a solution to the equation $f(x) = 0$. Divide the polynomial $f(x)$ by the polynomial $x + 1/2$:

	8	18	15	14
-1/2	8	14	8	10

Since the remainder of the division is 10, $x = -1/2$ is not a solution to the equation $f(x) = 0$. Divide the polynomial $f(x)$ by the polynomial $x + 1$:

	8	18	15	14
-1	8	10	5	9

Since the remainder of the division is 9, $x = -1$ is not a solution to the equation $f(x) = 0$. Divide the polynomial $f(x)$ by the polynomial $x + 7$:

	8	18	15	14
-7	8	-38	281	-1953

Since the remainder of the division is -1953, $x = -7$ is not a solution to the equation $f(x) = 0$. Divide the polynomial $f(x)$ by the polynomial $x + 7/2$:

	8	18	15	14
$-7/2$	8	-10	50	-161

Since the remainder of the division is -161, $x = -7/2$ is not a solution to the equation $f(x) = 0$. Divide the polynomial $f(x)$ by the polynomial $x + 7/4$:

	8	18	15	14
$-7/4$	8	4	8	0

In this case, the remainder is zero; therefore, the number $x = -7/4$ is the root of the equation $f(x) = 0$. Moreover, we get the decomposition

$$8x^3 + 18x^2 + 15x + 14 = (x + 7/4)(8x^2 + 4x + 8).$$

Since the equation $8x^2 + 4x + 8 = 0$ has no real solutions, $x = -7/4$ is the only root of the equation $f(x) = 0$. It remains to substitute the value of $x = -7/4$ into the second equation:

$$(10 + 4x)^y - 2 = y(5 + 7/x) + 7^{x+y} \cdot \sqrt{16x(x+1)^2 + 40x^2 + 89x + 49}.$$

After substitution, we get

$$3^y - 2 = y.$$

Let us consider the function $g(y) = 3^y - 2 - y$. One solution, $y = 1$, evident; it remains to check the existence of the second solution. The following inequalities are valid:

$$g(0) = -1 < 0,$$

$$g(-2) = 1/9 > 0,$$

from which the Weierstrass intermediate value theorem for a continuous function implies the existence of a value $y^* \in (-2; 0)$ such that $g(y^*) = 0$.

Thus, it is proved that the original system has at least two different solutions: $(x; y) = (-7/4; 1)$ and $(x; y) = (-7/4; y^*)$. $\qquad\square$

2.3   Reducing the Degree of the Equation

Sometimes it is possible to reduce the degree of the equation by successfully grouping its members and using known identities.

Example 2.4. Solve the equation

$$x^4 - 4x^3 - 4x^2 - 64 = 0. \qquad (2.7)$$

Proof. The equation can be solved using the method of root selection described above. It can be noted that

$$x^4 - 4x^3 - 4x^2 = x^2(x^2 - 4x + 4) = (x(x-2))^2,$$

and using the equality $a^2 - b^2 = (a - b)(a + b)$, get the equation

$$((x(x2))^2 - 64 = 0 \Rightarrow (x(x-2) - 8)(x(x-2) + 8) = 0$$

$$\Rightarrow (x^2 - 2x - 8)(x^2 - 2x + 8) = 0 \Rightarrow (x+2)(x-4)(x^2 - 2x + 8) = 0.$$

So, $x = -2$ and $x = 4$ because the equation $x^2 - 2x + 8 = 0$ has no roots. $\qquad\square$

Example 2.5. Solve the equation

$$(x + 2)(x + 4)(x + 6)(x + 8) = 105.$$

Proof. If we just open the brackets, we get an equation of the fourth degree with large coefficients, and the selection of roots will be difficult. However, you can note the equalities

$$(x + 2)(x + 8) = x^2 + 10x + 16,$$

$$(x + 4)(x + 6) = x^2 + 10x + 24.$$

This allows denoting $t = x^2 + 8x + 16$ to obtain the equation

$$t(t + 8) = 105, \quad t^2 + 8t - 105 = 0.$$

The roots of this equation are the numbers $t_1 = 7$ and $t_2 = -15$. Returning to the variable x, we get the equations

$$x^2 + 10x + 16 = 7 \Rightarrow x^2 + 10x + 9 = 0,$$

$$x^2 + 10x + 16 = -15 \Rightarrow x^2 + 10x + 31 = 0.$$

The first of them has the roots $x_1 = -1$ and $x_2 = -9$, whereas the second has no roots. $\qquad\square$

Example 2.6. A similar technique, as before, is applicable to solving the equation

$$(x^2 + x + 16)(x^2 - 20x + 16) + 54x^2 = 0.$$

Proof. Here, a hint of the approach to the solution is the presence of the same values $x^2 + 16$ in both brackets. A few interfere with x in the first bracket and $-20x$ in the second bracket. But we still have $54x^2$ in the equation. Note that $x = 0$ is not the root of the original equation. Therefore, when dividing both parts by x^2, we get the equation

$$\left(\frac{x^2 + x + 16}{x}\right)\left(\frac{x^2 - 20x + 16}{x}\right) + 54 = 0,$$

equivalent to the original the equation, i.e., the equation

$$\left(x + \frac{16}{x} + 1\right)\left(x + \frac{16}{x} - 20\right) + 54 = 0.$$

Denoting $x + \frac{16}{x} = t$, we obtain the equation

$$(t + 1)(t - 20) + 54 = 0 \Rightarrow t^2 - 19t + 34 = 0,$$

from which we get $t_1 = 2$ and $t_2 = 17$. Going back to x,

$$x + \frac{16}{x} = 2, \quad x + \frac{16}{x} = 17.$$

The first equation (given that $x \neq 0$) is transformed to the form

$$x^2 - 2x + 16 = 0.$$

It has no solutions. And the second equation is reduced to the form

$$x^2 - 17x + 16 = 0,$$

and its roots are $x_1 = 1$ and $x_2 = 16$.   □

Example 2.7. Solve the equation

$$x^4 - 4x^3 + 6x^2 - 4x + 1 = 0.$$

Proof. You can immediately note that $x = 1$ is the root of this equation and use the techniques described above. But here, we use a technique similar to that used in solving the previous problem. Noting that $x = 0$ is not the root of this equation, we divide both its parts by x^2 and get the equation

$$x^2 - 4x + 6 - \frac{4}{x} + \frac{1}{x^2} = 0,$$

equivalent to the original. Denote $t = x + \frac{1}{x}$, and note that

$$t^2 = \left(x + \frac{1}{x}\right)^2 = x^2 + 2 + \frac{1}{x^2} \implies x^2 + \frac{1}{x^2} = t^2 - 2.$$

Therefore, our equation will take the form

$$t^2 - 2 - 4t + 6 = 0 \Rightarrow t^2 - 4t + 4 = 0,$$

which gives $t = 2$. Returning to the variable x, we get

$$x + \frac{1}{x} = 2 \Rightarrow \frac{(x-1)^2}{x} = 0 \Rightarrow x = 1.$$

This equation is a special case of the so-called return equation.   □

Example 2.8. Solve the equation

$$4x^2(2x+1)^2 - 2x(4x^2 - 1) = 30(2x - 1)^2.$$

Proof. This equation belongs to the so-called homogeneous equations because it can be represented as

$$(2x(2x+1))^2 - (2x(2x+1))(2x - 1) = 30(2x - 1)^2,$$

so

$$u^2(x) - u(x)v(x) - 30v^2(x) = 0,$$

where $u(x) = 2x(2x+1)$, $v(x) = 2x - 1$. The left side of this equation can be represented as

$$(u(x) - 6v(x))(u(x) + 5v(x)),$$

and the equation considered is equivalent to the set of equations

$$u(x) = 6v(x) \quad \text{and} \quad u(x) = -5v(x),$$

or

$$2x(2x+1) = 6(2x-1) \quad \text{and} \quad 2x(2x+1) = -5(2x-1),$$

so

$$4x^2 - 10x + 6 = 0 \quad \text{or} \quad 4x^2 + 12x - 5 = 0,$$

which gives

$$x_1 = 1, \quad x_2 = \frac{3}{2}, \quad x_3 = \frac{-3 + \sqrt{14}}{2}, \quad x_4 = \frac{-3 - \sqrt{14}}{2}. \qquad \square$$

Let us describe another technique that is useful when searching for the factorization of a polynomial.

Example 2.9. Solve the equation

$$x^4 + 2x^3 - 3x^2 - 4ax - a^2 = 0.$$

Proof. Let us solve it in two ways.

1. In this problem, it is relatively easy to note that

$$x^4 + 2x^3 - 3x^2 = (x(x+1))^2 - 4x^2,$$

and the equation will take the form

$$(x(x+1))^2 - 4x^2 - 4ax - a^2 = 0 \Rightarrow (x(x+1))^2 - (2x+a)^2 = 0$$
$$\Rightarrow (x^2 + x - 2x - a)(x^2 + x + 2x + a) = 0$$
$$\Rightarrow (x^2 - x - a)(x^2 + 3x + a) = 0.$$

The equation

$$x^2 - x - a = 0$$

has the roots

$$x_1 = \frac{1 + \sqrt{1 + 4a}}{2}, \quad x_2 = \frac{1 - \sqrt{1 - 4a}}{2}$$

if $a \geq -\frac{1}{4}$. The second equation has the roots

$$x_3 = \frac{-3 + \sqrt{9 - 4a}}{2}, \quad x_4 = \frac{-3 - \sqrt{9 - 4a}}{2}$$

if $a \leq \frac{9}{4}$.

2. Let us propose another solution to this problem, in which you do not need to note anything, but you can solve using standard methods. The original equation can be considered a quadratic equation with respect to a, the coefficients of which depend on x:

$$a^2 + 4ax - x^4 - 2x^3 + 3x^2 = 0.$$

Its discriminant is equal to

$$16x^2 + 4x^4 + 8x^3 - 12x^2 = 4x^4 + 8x^3 + 4x^2 = 4x^2(x^2 + 2x + 1) = 4x^2(x+1)^2.$$

Therefore, for all x, this equation is equivalent to the set of equations

$$a_1 = \frac{-4x + 2x(x+1)}{2} = x^2 - x,$$

$$a_2 = \frac{-4x - 2x(x+1)}{2} = -x^2 - 3x.$$

It remains to consider the obtained equalities as equations with respect to x. This has already been done in the first solution of the original equation.   □

2.4 Using Function Properties

When solving algebraic equations, approaches based on the study of the monotony of the functions under consideration on finding the largest or smallest values of the quantities included in them are useful.

Example 2.10. Solve the equation

$$x^{2023} + 2022x = 2023.$$

Proof. There is an increasing function on the left side, so the equation cannot have more than one solution. It is immediately clear that $x = 1$ is the root of this equation. According to the above, there are no other roots. □

Example 2.11. Solve the equation

$$x^4 - 17x^2 - 6x + 90 = 0.$$

Proof. Of course, we can search for integer roots (and they will be found), but we can do the non-obvious transformation of the equation to the form

$$x^4 - 18x^2 + 81 + x^2 - 6x + 9 = 0 \Rightarrow (x^2 - 9)^2 + (x - 3)^2 = 0.$$

From this, we immediately find the answer: $x = 3$. □

The following equation, more complex at first sight, is based on the same idea.

Example 2.12. Solve the equation

$$(x^4 - 8x^2 + 18)^2 + (x^2 + 4x + 7)^2 = 13.$$

Proof. To solve it, we transform the left part to the form

$$((x^2 - 4)^2 + 2)^2 + ((x + 2)^2 + 3)^2 = 13.$$

The smallest value accepted by the expression

$$((x^2 - 4)^2 + 2)^2$$

is equal to $2^2 = 4$, and it is achieved at $x^2 = 4 \Rightarrow x = \pm 2$. The smallest value used by the expression $((x+2)^2+3)^2$ is equal to 9, it is accepted when $x = -2$. Therefore, the number 13 is the smallest value of the function on the left side of the equation, and it is reached at $x = -2$. □

The following is another example of using function properties.

Example 2.13. For all a, solve the equation

$$32x^{10} + (3x + a)^5 + 10x^2 + 15x + 5a = 0.$$

Proof. The hint of a solution is contained in the equality

$$15x + 5a = 5(3x + a).$$

Then, the original equation can be rewritten as

$$32x^{10} + 10x^2 + (3x + a)^5 + 5(3x + a) = 0,$$

or

$$(2x^2)^5 + 5(2x^2) + (3x + a)^5 + 5(3x + a) = 0.$$

Let us put $f(t) = t^5 + 5t$. Then, the equation will take the form

$$f(2x^2) + f(3x + a) = 0 \Longrightarrow f(2x^2) = -f(3x + a).$$

But $-f(z) = f(-z)$, so we get

$$f(2x^2) = f(-3x - a).$$

Since $f(z)$ increases, this is equivalent to the equation

$$2x^2 = -3x - a \Rightarrow 2x^2 + 3x + a = 0.$$

Its discriminant is $D(a) = 9 - 8a$.

- If $a > \frac{9}{8}$, then there are no solutions.
- When $a = \frac{9}{8}$, the only root is $x = -\frac{3}{4}$.
- With $a < \frac{9}{8}$, we get

$$x_1 = \frac{-3 + \sqrt{9 - 8x}}{4}, \quad x_2 = \frac{-3 - \sqrt{9 - 8x}}{4}.$$ □

2.5   Trigonometric Substitutions of Variables

Example 2.14. Solve the system

$$\begin{cases} 8x(2x^2 - 1)(8x^4 - 8x^2 + 1) = 1, \\ 0 \le x \le 1. \end{cases}$$

Proof. We can solve this problem by replacing the variable $x = \cos t$, $0 \le t \le \frac{\pi}{2}$.

Since

$$2\cos^2 t - 1 = \cos 2t, \quad 8\cos^4 t - 8\cos^2 t + 1 = \cos 4t,$$

the first equation takes the form

$$8\cos t \cos 2t \cos 4t = 1.$$

This is a fairly well-known trigonometric equation, and its solution is also discussed in the chapter on trigonometry. Note first that t such that $\sin t = 0$ is not the root of this equation (if $\sin t = 0$, the left-hand side is not 0). Therefore, the equation

$$8\sin t \cos t \cos 2t \cos 4t = \sin t \tag{2.8}$$

has an extra root, $t = 0$, compared to the original one. But the equation is easy to solve by noting that

$$8\sin t \cos t = 4\sin 2t,$$

$$4\sin 2t \cos 2t = 2\sin 4t,$$

$$2\sin 4t \cos 4t = \sin 8t,$$

and then equation (2.8) is equivalent to the equation $\sin 8t = \sin t$. Using the trigonometric formula $\sin \alpha - \sin \beta = 2\cos \frac{\alpha+\beta}{2} \sin \frac{\alpha-\beta}{2}$, we get that our equation is equivalent to the equation

$$2\cos \frac{9t}{2} \sin \frac{7t}{2} = 0.$$

The equation

$$\cos \frac{9t}{2} = 0$$

has the roots $t = \frac{\pi}{9} + \frac{2}{9}\pi k$ and $k \in \mathbb{Z}$. The condition $0 \le t \le \pi$ is satisfied by the numbers $\frac{\pi}{9}$ and $\frac{\pi}{3}$. The equation

$$\sin \frac{7t}{2} = 0$$

has the roots $t = \frac{2}{7}\pi k$ and $k \in \mathbb{Z}$. The condition $0 \le t \le \frac{\pi}{2}$ is satisfied by $t = 0$ and $t = \frac{2\pi}{7}$. But, as noted above, $t = 0$ is not the root of the equation.

Answer: $x = \cos \frac{\pi}{9}$, $x = \frac{1}{2}$, $x = \cos \frac{2\pi}{7}$. $\square$

2.6   Solving Inequalities

Rational inequalities can be reduced to one of the following types:

$$\frac{P(x)}{Q(x)} > 0, \quad \frac{P(x)}{Q(x)} < 0, \quad \frac{P(x)}{Q(x)} \ge 0, \quad \frac{P(x)}{Q(x)} \le 0,$$

where $P(x)$, $Q(x)$ are polynomials. According to a well-known theorem, any polynomial

$$P(x) = a_n x^n + a_{n-1} x^{n-1} + \cdots + a_1 x + a_0, \quad a_n \ne 0$$

of degree n can be represented as

$$P(x) = a_n (x - \alpha_1)^{k_1} \ldots (x - \alpha_r)^{k_r} (x^2 + p_1 x + q_1)^{l_1} (x^2 + p_s x + q_s)^{l_s},$$

where $k_1 + \cdots + k_r + 2(l_1 + \cdots + l_s) = n$ and the polynomials $x^2 + p_i x + q_i$, $i = 1, \ldots, s$ have no roots. Let us suppose that the denominator of the fraction $Q(x) = b_m x^m + \cdots + b_1 x + b_0$, $b_0 \ne 0$, is of the form

$$Q(x) = b_m (x - \beta_1)^{g_1} \ldots (x - \beta_u)^{g_u} (x^2 + c_1 x + d_1)^{f_1} \ldots (x^2 + c_v x + d_v)^{f_v},$$

where $g_1 + \cdots + g_u + 2(f_1 + \cdots + f_v) = m$ and the polynomials $x^2 + c_j x + d_j$, $j = 1, \ldots, v$ have no roots. Since for all x polynomials

$$x^2 + p_i x + q_i, \quad i = 1, \ldots, s \quad \text{and} \quad x^2 + c_j x + d_j, \quad j = 1, \ldots, v$$

are positive, the initial inequality, for example,

$$\frac{P(x)}{Q(x)} \ge 0,$$

can be replaced by an inequality equivalent to it:

$$\frac{a_n}{b_m} \frac{(x - \alpha_1)^{k_1} \ldots (x - \alpha_r)^{k_r}}{(x - \beta_1)^{g_1} \ldots (x - \beta_u)^{g_u}} \ge 0.$$

Let $\gamma_1,\ldots,\gamma_t$ be the numbers $\alpha_1,\ldots,\alpha_r$ and $\beta_1,\ldots,\beta_u$, arranged in ascending order. We assume that $\frac{a_n}{b_m} > 0$ (otherwise, we simply consider the inequality $-\frac{P(x)}{Q(x)} \leq 0$). If it turns out that some of the numbers α_i coincide with the numbers β_j, then the number corresponding to these numbers γ_l is excluded from the definition of inequality, and the expression

$$\frac{(x-\alpha_i)^{k_i}}{(x-\beta_j)^{g_j}}$$

is replaced by

$$(x-\gamma_l)^{k_i-g_j} = (x-\gamma_l)^{h_l}.$$

Thus, the initial inequality is equivalent to the inequality

$$(x-\gamma_1)^{h_1}\ldots(x-\gamma_t)^{h_t} \geq 0.$$

- If the number $h_i \leq 0$, then the point γ_i is excluded from the definition area. It remains to be noted that if $x > \gamma_t$, then all brackets are positive, and the inequality is satisfied.
- If h_t is an even number, then on the interval $(\gamma_{t-1};\gamma_t)$, the bracket $(x-\gamma_t)^{h_t} > 0$. Therefore, this interval will be included in the set of solutions.
- If h_t — is an odd number, then on the interval $(\gamma_{t-1};\gamma_t)$, the bracket $(x-\gamma_t)^{h_t}$ is negative and with it the entire product of the brackets on the left side of the inequality.

Continuing this movement to the left, we find the signs of the function in question at all intervals $(\gamma_{i-1};\gamma_i)$, and thereby, we obtain a solution to the inequality. This process is called the interval method.

Example 2.15. Solve the inequality

$$\frac{x^2-3x+24}{x^2-3x+3} < 4.$$

Proof. Convert it to the form

$$\frac{x^2-3x+24-4x^2+12x-12}{x^2-3x+3} < 0 \implies \frac{-3x^2+9x+12}{x^2-3x+3} < 0$$

$$\implies \frac{x^2-3x-4}{x^2-3x+3} > 0.$$

Further,

$$x^2-3x-4 = (x+1)(x-4),$$

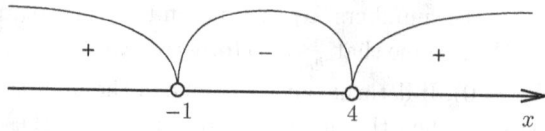

Fig. 2.1.   Signs of $(x+1)(x-4)$.

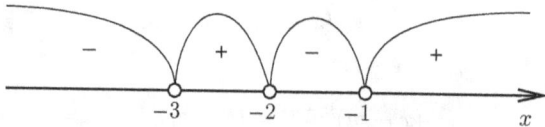

Fig. 2.2.   Signs of $\frac{1}{(x+1)(x+2)(x+3)}$.

and the polynomial $x^2 - 3x + 3$ has no roots. Therefore, the original inequality is equivalent to the inequality $(x+1)(x-4) > 0$. Signs of expression $(x+1)(x-4)$ in Fig. 2.1.

The solutions are $x \in (-\infty; -1) \cup (4; +\infty)$. ☐

Example 2.16. Solve the inequality

$$\frac{(x-1)(x-2)(x-3)}{(x+1)(x+2)(x+3)} > 1.$$

Proof. Transform the inequality to the form

$$\frac{(x-1)(x-2)(x-3) - (x+1)(x+2)(x+3)}{(x+1)(x+2)(x+3)} > 0$$

$$\Rightarrow \frac{-6(x^2+1)}{(x+1)(x+2)(x+3)} > 0 \Rightarrow \frac{1}{(x+1)(x+2)(x+3)} < 0.$$

Placing the signs as the real axis (see in Fig. 2.2).

From this, we get the solution.

Answer: $x \in (-\infty; -3) \cup (-2; -1)$. ☐

Example 2.17. Find all values of a for which the inequality

$$\frac{x - 3a - 1}{x + 2a - 2} \leq 0$$

holds, for all x, on the interval $[2; 3]$.

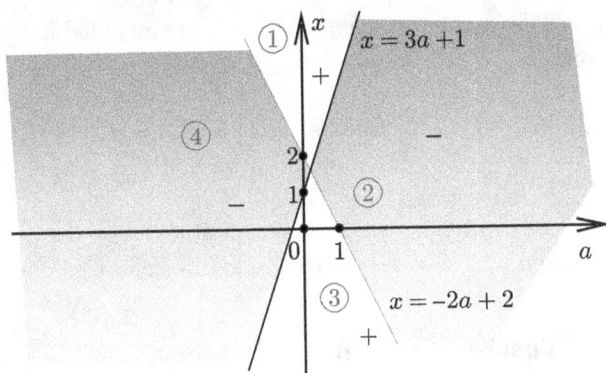

Fig. 2.3.   Signs of $\frac{x-3a-1}{x+2a-2}$.

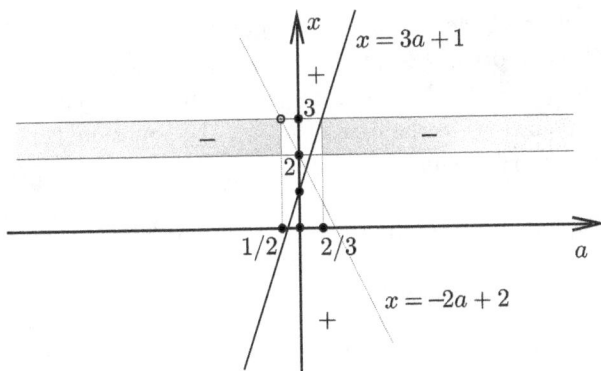

Fig. 2.4.   The values of a specifying the answer.

Proof. It is convenient to use the area method here (which will be discussed in detail later), similar to the interval method described above. To do this, consider a plane. Note that above the line $x = 3a+1$, the function $x-3a-1$ is positive, and under this line, it is negative. Similarly, above the line $x = -2a + 2$, the expression $x + 2a - 2$ is positive, and below it, the expression is negative. This immediately tells us that the whole fraction is positive in the areas ① and ③ and negative in the areas ② and ④ in the Fig. 2.3.

The set of points $x = 3a+1$ is included in the set of solutions, and the set of points $x = -2a+2$ is excluded from the set of solutions. The solution to the problem are those a for which all $x \in [2; 3]$ are included in the resulting set of solutions

- Equating $x = 3a + 1$ to the number 3, we find $a = \frac{2}{3}$.
- Equating $x = -2a + 2$ to the number 3, we find $a = -\frac{1}{2}$.

As a result see in Fig. 2.4, we get $a \geq \frac{2}{3}$ and $a < -\frac{1}{2}$ (the point $a = -\frac{1}{2}$ is excluded since $a = -\frac{1}{2}x = 3$ is not included in the set of solutions). □

2.7   Vieta's Theorem for High-Order Equations

Definition 2.3. Consider again a function of the form

$$f(x) = a_n x^n + a_{n-1} x^{n-1} + \cdots + a_1 x + a_0, \quad n \in \mathbb{N}, \quad a_n \neq 0, \quad a_k \in \mathbb{R}, \tag{2.9}$$

which we call *an algebraic polynomial of degree n*. In the case of $n = 2$, the equation $f(x) = 0$ will be called *quadratic*, or a second-order equation. In the case of $n = 3$, the equation $f(x) = 0$ will be called *cubic*, or a third-order equation. In the case of $n = 4$, the equation $f(x) = 0$ will be called a fourth-order equation.

Let $x_1, x_2, \ldots, x_n$ — be the roots of equation (2.9). Then,

$$a_n x^n + a_{n-1} x^{n-1} + \cdots + a_1 x + a_0 = 0,$$

$$a_n \neq 0 \Longleftrightarrow a_n(x - x_1)(x - x_2) \ldots (x - x_n) = 0.$$

Here, by opening the brackets and equating the coefficients with the same degrees, we arrive at the following formulas:

$$x_1 + x_2 + \cdots + x_n = -a_{n-1}/a_n,$$

$$(x_1 x_2 + x_1 x_3 + x_1 x_4 + \cdots + x_1 x_n) + (x_2 x_3 + x_2 x_4 + \cdots + x_2 x_n)$$

$$+ \cdots + x_{n-1} x_n = a_{n-2}/a_n,$$

$$\cdots \cdots \cdots \cdots \cdots \cdots \cdots \cdots \cdots \cdots \cdots \cdots \cdots \cdots$$

$$\sum_{i_1 < i_2 < \cdots < i_k} x_{i_1} x_{i_2} \ldots x_{i_k} = (-1)^k a_{n-k}/a_n,$$

$$\cdots \cdots \cdots \cdots \cdots \cdots \cdots \cdots \cdots \cdots \cdots \cdots \cdots \cdots$$

$$x_1 x_2 \ldots x_n = (-1)^n a_0/a_n,$$

which are called *Vieta's formulas*.

Remark 2.3. We give explicit formulas for the case of polynomials of degrees 2, 3, and 4.

1. For the quadratic equation $a_2x^2 + a_1x + a_0 = 0$, $a_2 \neq 0$, where x_1, x_2 are the roots of the equation (the case of $D \geq 0$), the general formulas take the form

$$\begin{cases} x_1 + x_2 = -a_1/a_2, \\ x_1x_2 = a_0/a_2. \end{cases}$$

2. For the cubic equation $a_3x^3 + a_2x^2 + a_1x + a_0 = 0$, $a_3 \neq 0$, let x_1, x_2, x_3 be the roots of the equation. The general formulas take the form

$$\begin{cases} x_1 + x_2 + x_3 = -a_2/a_3, \\ x_1x_2 + x_1x_3 + x_2x_3 = a_1/a_3, \\ x_1x_2x_3 = -a_0/a_3. \end{cases}$$

3. For a fourth degree equation $a_4x^4 + a_3x^3 + a_2x^2 + a_1x + a_0 = 0$, $a_4 \neq 0$, let x_1, x_2, x_3, x_4 be the roots of the equation. The general formulas take the form

$$\begin{cases} x_1 + x_2 + x_3 + x_4 = -a_3/a_4, \\ x_1x_2 + x_1x_3 + x_1x_4 + x_2x_3 + x_2x_4 + x_3x_4 = a_2/a_4, \\ x_1x_2x_3 + x_1x_2x_4 + x_2x_3x_4 = -a_1/a_4, \\ x_1x_2x_3x_4 = a_0/a_4. \end{cases}$$

Example 2.18. Determine all the values of a for each of which there are three different roots of the equation

$$x^3 + (a^2 - 9a)x^2 + 8ax - 64 = 0$$

which form a geometric progression. Find these roots.

Proof. We solve it in two ways: with and without using Vieta's theorem.

1. Let q be the denominator of the geometric progression. Then, the roots are connected by the relation $x_2 = qx_1$, $x_3 = q^2x_1$. Using Vieta's theorem,

$x_1 x_2 x_3 = 64$ or $(q x_1)^3 = 64$, from which we have $x_2 = 4$. Let's write down Vieta's theorem for

$$x_1 = q^{-1} x_2 = 4 q^{-1}, \quad x_2 = 4, \quad x_3 = q x_2 = 4q.$$

$$\begin{cases} 4(q^{-1} + 1 + q) = -(a^2 - 9a), \\ 16(q + 1 + q^{-1}) = 8a, \\ x_2 = 4. \end{cases}$$

Note that $a \neq 0$ because otherwise the equation $q + 1 + q^{-1} = 0$ has no solutions; therefore, this case contradicts the condition of the existence of three different roots. Dividing the first equation by the second, we get

$$2 = -(a - 9) \Longleftrightarrow a = 7.$$

Thus, from the second equation, we find

$$q + 1 + q^{-1} = 7/2 \Longleftrightarrow q^2 - (5/2)q + 1 = 0 \Longleftrightarrow q = 2, \quad q = 1/2.$$

Let $q = 2$. Then, $x_1 = 2$, $x_2 = 4$, and $x_3 = 8$. Let $q = 1/2$. Then, $x_1 = 8$, $x_2 = 4$, and $x_3 = 2$. In both cases, we get that the roots of the original equation are the numbers 2, 4, and 8.

2. Finding, as before, the root $x_2 = 4$, and substituting it into the original equation, we get

$$4^3 + 16(a^2 - 9a) + 32a - 64 = 0$$
$$\Longleftrightarrow a^2 - 7a = 0.$$

We get two values of the parameter a: $a = 0$ and $a = 7$. As before, we prove that the case of $a = 0$ is impossible. For $a = 7$, the equation takes the form

$$x^3 - 14x + 56x - 64 = 0 \Longleftrightarrow (x - 4)(x^2 - 10x + 16) = 0$$
$$\Longleftrightarrow (x - 4)(x - 2)(x - 8) = 0.$$

Therefore, we get the desired answer.

Answer: $a = 7$, the roots of the equation are 2, 4, 8. $\square$

2.7.1 Discriminant of a polynomial of degree n

Definition 2.4. The *discriminant* of the equation

$$f(x) = a_n x^n + a_{n-1} x^{n-1} + \cdots + a_1 x + a_0, \quad n \in \mathbb{N}, \quad a_n \neq 0, \quad a_k \in \mathbb{R},$$

is defined by the formula

$$D = a_n^{2n-2} \prod_{k>l} (x_k - x_l)^2,$$

where $x_1, x_2, \ldots, x_n$ are the roots of the equation $f(x) = 0$.

In the case of $n = 2$, according to Vieta's formulas, the discriminant of the equation $ax^2 + bx + c = 0$, $a \neq 0$, takes the form

$$D = a^2(x_1 - x_2)^2 = a^2((x_1 + x_2)^2 - 4x_1 x_2)$$

$$= a^2 \left(\left(-\frac{b}{a} \right)^2 - 4 \left(\frac{c}{a} \right) \right) = b^2 - 4ac.$$

We have obtained a well-known formula for the discriminant of a quadratic equation.

In the case of $n = 3$, we find the discriminant of the cubic equation $x^3 + px + q = 0$, $p, q \in \mathbb{R}$. The following equality is true:

$$(x_1^2 + x_2^2 - 2x_1 x_2)(x_1^2 + x_3^2 - 2x_1 x_3)(x_2^2 + x_3^2 - 2x_2 x_3)$$

$$= -4(x_1 x_2 + x_1 x_3 + x_2 x_3)^3 - 27(x_1 x_2 x_3 x_4)^2.$$

Now, let x_1, x_2, x_3 be the roots of the equation $x^3 + px + q = 0$. Then, from the preceding equality and Vieta's formulas, it follows that

$$D = (x_1 - x_2)^2 (x_1 - x_3)^2 (x_2 - x_3)^2 = -4p^3 - 27q^2$$

$$= -108 \cdot \left(\left(\frac{p}{3} \right)^3 + \left(\frac{q}{2} \right)^2 \right). \tag{2.10}$$

The discriminant of the cubic equation in the general case

$$a_3 x^3 + a_2 x^2 + a_1 x + a_0 = 0, \quad a_3 \neq 0, \quad a_3, a_2, a_1, a_0 \in \mathbb{R},$$

takes a more complex form:

$$D = a_1^2 a_2^2 - 4a_0 a_2^3 - 4a_1^3 a_3 + 18a_0 a_1 a_2 a_3 - 27a_0^2 a_3^2. \qquad (2.11)$$

2.8   Cardano Formulas

The equations of the third degree $ax^3 + bx^2 + cx + d = 0$, $a \neq 0$, can be easily reduced to an equation of the form $x^3 + px + q = 0$ (it is enough to divide by a first and then make the substitution $y = x + a_1/3$, where $a_1 = b/a$).

For the equation $x^3 + px + q = 0$, we introduce the[1] notation

$$\varepsilon = \frac{-1 + \sqrt{3}i}{2} = e^{2\pi i/3};$$

$$u = \sqrt[3]{-\frac{q}{2} + \sqrt{\left(\frac{p}{3}\right)^3 + \left(\frac{q}{2}\right)^2}};$$

$$v = \sqrt[3]{-\frac{q}{2} - \sqrt{\left(\frac{p}{3}\right)^3 + \left(\frac{q}{2}\right)^2}}.$$

Let x_1, x_2, x_3 be the roots of the equation $x^3 + px + q = 0$. Denote

$$z_1 := x_1 + \varepsilon x_2 + \varepsilon^{-1} x_3, \quad z_2 := x_1 + \varepsilon^{-1} x_2 + \varepsilon x_3.$$

Let us check the validity of the following equalities:

$$z_1^3 + z_2^3 = -27q, \quad z_1 z_2 = -3p.$$

Indeed, with the help of the equality

$$\varepsilon^3 = (\varepsilon^{-1})^3 = 1, \quad \varepsilon + \varepsilon^{-1} = -1,$$

$$\varepsilon^2 + \varepsilon^{-2} = (\varepsilon + \varepsilon^{-1})^2 - 2\varepsilon \cdot \varepsilon^{-1} = 1 - 2 = -1,$$

and Vieta's theorem

$$x_1 + x_2 + x_3 = 0, \quad x_1 x_2 x_3 = -q,$$

---

[1]By u and v, we mean arbitrarily chosen fixed values of the corresponding roots. This will further facilitate the calculation of the roots of the cubic equation by the Cardano formulas.

we get

$$z_1^3 + z_2^3 = x_1^3 + x_2^3 + x_3^3 + 3\varepsilon x_1 x_2(x_1 + \varepsilon x_2) + 3\varepsilon^{-1} x_1 x_3(x_1 + \varepsilon^{-1} x_3)$$

$$+ 3x_2 x_3(\varepsilon x_2 + \varepsilon^{-1} x_3) + 6x_1 x_2 x_3 + x_1^3 + x_2^3 + x_3^3$$

$$+ 3\varepsilon^{-1} x_1 x_2(x_1 + \varepsilon^{-1} x_2) + 3\varepsilon x_1 x_3(x_1 + \varepsilon x_3)$$

$$+ 3x_2 x_3(\varepsilon^{-1} x_2 + \varepsilon x_3) + 6x_1 x_2 x_3$$

$$= 2(x_1^3 + x_2^3 + x_3^3) - 3x_1 x_2(x_1 + x_2) - 3x_1 x_3(x_1 + x_3)$$

$$- 3x_2 x_3(x_2 + x_3) + 12x_1 x_2 x_3$$

$$= 2(x_1 + x_2 + x_3)^3 - 9x_1 x_2(x_1 + x_2)$$

$$- 9x_1 x_3(x_1 + x_3) - 9x_2 x_3(x_2 + x_3)$$

$$= 2 \cdot 0 - 9x_1 x_2(-x_3) - 9x_1 x_3(-x_2) - 9x_2 x_3(-x_1)$$

$$= 27x_1 x_2 x_3 = -27q,$$

and

$$z_1 z_2 = x_1^2 + x_2^2 + x_3^2 - (x_1 x_2 + x_1 x_3 + x_2 x_3)$$

$$= (x_1 + x_2 + x_3)^2 - 3(x_1 x_2 + x_1 x_3 + x_2 x_3)$$

$$= 0 - 3p = -3p.$$

Thus, it remains to conclude that z_1^3 and z_2^3 are the roots of the quadratic equation

$$y^2 + 27qy - 27p^3 = 0.$$

Let us solve this equation (for example, using the discriminant of the quadratic equation) and find

$$y_{1,2} = \frac{-27q \pm \sqrt{27^2 q^2 + 4 \cdot 27p^3}}{2} = 27 \cdot \left(-\frac{q}{2} \pm \sqrt{\left(\frac{p}{3}\right)^3 + \left(\frac{q}{2}\right)^2} \right).$$

Hence,

$$z_1 = 3v, \qquad z_2 = 3u.$$

We obtain the system

$$\begin{cases} x_1 + x_2 + x_3 = 0, \\ x_1 + \varepsilon x_2 + \varepsilon^{-1} x_3 = 3v, \\ x_1 + \varepsilon^{-1} x_2 + \varepsilon x_3 = 3u. \end{cases}$$

Adding up all the equalities, we arrive at the formula $x_1 = u + v$. Multiply the second equation in the system by ε^{-1} and the third by ε and sum up the resulting three equations. As a result, we get $x_2 = \varepsilon u + \varepsilon^{-1} v$. Now, multiply the second equation in the system by ε and the third by ε^{-1} and sum up the resulting three equations. As a result, we get $x_3 = \varepsilon^{-1} u + \varepsilon v$. Thus, the Cardano formulas[2] are obtained:

$$x_1 = u + v;$$

$$x_2 = \varepsilon u + \varepsilon^{-1} v;$$

$$x_3 = \varepsilon^{-1} u + \varepsilon v.$$

Example 2.19. Solve the equation

$$x^3 - 3x = 0.$$

Proof. In this case, it is easy to make sure that $x_1 = 0$ and $x_2 = -\sqrt{3}$, $x_3 = \sqrt{3}$ are the roots of the equation. We get them using the Cardano formulas using the fact that one of the roots of $\sqrt[6]{-1}$ is i. In fact,[3] $i^6 = -1$. In our case, we get

$$p = -3, \quad q = 0;$$

$$\varepsilon = \frac{-1 + \sqrt{3}i}{2} = e^{2\pi i/3}, \quad \varepsilon^{-1} = \frac{-1 - \sqrt{3}i}{2} = e^{-2\pi i/3};$$

$$u = \sqrt[3]{0 + \sqrt{-1 + 0}} = \sqrt[6]{-1} = i;$$

$$v = \sqrt[3]{-0 - \sqrt{-1 + 0}} = -\sqrt[6]{-1} = -i.$$

---

[2]If the roots are understood as multivalued functions, then the Cardano formulas can be written as

$$x_{1,2,3} = \sqrt[3]{-\frac{q}{2} + \sqrt{\left(\frac{p}{3}\right)^3 + \left(\frac{q}{2}\right)^2}} + \sqrt[3]{-\frac{q}{2} - \sqrt{\left(\frac{p}{3}\right)^3 + \left(\frac{q}{2}\right)^2}}.$$

And we use discriminant (2.10) in the form

$$x_{1,2,3} = \sqrt[3]{-\frac{q}{2} + \sqrt{-\frac{D}{108}}} + \sqrt[3]{-\frac{q}{2} - \sqrt{-\frac{D}{108}}}.$$

Note that although these formulas look more compact, they are cumbersome for specific calculations since they have to calculate the square root of a complex number and calculate twice the cubic roots of various complex numbers.

[3]Recall here that as u and v, we choose an arbitrary value of the corresponding root.

According to the Cardano formulas, we find

$$x_1 = u + v = 0;$$

$$x_2 = \varepsilon u + \varepsilon^{-1} v = e^{2\pi i/3} \cdot i - e^{-2\pi i/3} \cdot i = -\sqrt{3};$$

$$x_3 = \varepsilon^{-1} u + \varepsilon v = e^{-2\pi i/3} \cdot i - e^{2\pi i/3} \cdot i = \sqrt{3}. \qquad \square$$

Example 2.20. Solve the equation

$$x^3 - 3x + 2 = 0.$$

Proof. In this case, it is easy to make sure that $x_1 = -2$, $x_2 = 1$, and $x_3 = 1$ are the roots of the equation. For example, this follows from the representation $x^3 - 3x + 2 = (x - 1)^2 (x + 2)$. We find the roots using the Cardano formulas. We have

$$p = -3, \quad q = 2;$$

$$u = \sqrt[3]{-1 + \sqrt{-1 + 1}} = -1, \quad v = \sqrt[3]{-1 - \sqrt{-1 + 1}} = -1.$$

According to the Cardano formulas, we find

$$x_1 = u + v = -2;$$

$$x_2 = \varepsilon u + \varepsilon^{-1} v = -e^{2\pi i/3} - e^{-2\pi i/3} = 1;$$

$$x_3 = \varepsilon^{-1} u + \varepsilon v = -e^{-2\pi i/3} - e^{2\pi i/3} = 1. \qquad \square$$

Example 2.21. Solve the equation

$$x^3 + 3x = 0.$$

Proof. In this case, it is easy to make sure that $x_1 = 0$, $x_2 = \sqrt{3}i$, and $x_3 = -\sqrt{3}i$ are the roots of the equation. We get them using the Cardano formulas. We have

$$p = -3, \quad q = 0;$$

$$u = \sqrt[3]{0 + \sqrt{1 + 0}} = 1;$$

$$v = \sqrt[3]{-0 - \sqrt{1 + 0}} = 1.$$

According to the Cardano formulas, we find

$$x_1 = u + v = 0;$$

$$x_2 = \varepsilon u + \varepsilon^{-1} v = e^{2\pi i/3} - e^{-2\pi i/3} = \sqrt{3}i;$$

$$x_3 = \varepsilon^{-1} u + \varepsilon v = e^{-2\pi i/3} - e^{2\pi i/3} = -\sqrt{3}i. \qquad \square$$

Example 2.22. Solve the equation

$$x^3 - 3x^2 - 9x - 5 = 0.$$

Proof. Since the coefficient at x^2 is different from zero, you need to replace $y = x + b/3a = x - 1$. Therefore,

$$x^3 - 3x^2 - 9x - 5 = 0$$

$$\Longleftrightarrow (y+1)^3 - 3(y+1)^2 - 9(y+1) - 5 = 0$$

$$\Longleftrightarrow (y^3 + 3y^2 + 3y + 1) - 3(y^2 + 2y + 1) - 9(y+1) - 5 = 0$$

$$\Longleftrightarrow y^3 - 12y - 16 = 0.$$

Let us find the roots of the last equation using the Cardano formulas. We have

$$p = -12, \quad q = -16;$$

$$u = \sqrt[3]{8 + \sqrt{-64 + 64}} = -2, \quad v = \sqrt[3]{8 - \sqrt{-64 + 64}} = -2.$$

According to the Cardano formulas, we find

$$y_1 = u + v = 4;$$

$$x_2 = \varepsilon u + \varepsilon^{-1} v = 2(e^{2\pi i/3} + e^{-2\pi i/3}) = -2;$$

$$x_3 = \varepsilon^{-1} u + \varepsilon v = 2(e^{-2\pi i/3} + e^{2\pi i/3}) = -2.$$

By replacing $y = x - 1$, we get $x = y + 1$. Therefore, $x_1 = 5$, $x_2 = -1$, and $x_3 = -1$ are roots of the original equation.

Indeed, it is easy to make sure that $x_1 = 5$, $x_2 = -1$, and $x_3 = -1$ are roots of the equation. For example, this follows from the representation $x^3 - 3x - 9x - 5 = (x+1)^2(x-5)$. $\qquad \square$

2.8.1 *Discriminant of the cubic equation*

Let us study the discriminant in more detail. Let x_1, x_2, x_3 be the roots of the equation $a_3 x^3 + a_2 x^2 + a_1 x + a_0 = 0$, $a_3 \neq 0$, $a_3, a_2, a_1, a_0 \in \mathbb{R}$. For a polynomial with real coefficients, if there is a complex root, then the complex-conjugate one will necessarily be the root of the equation. Therefore, there are exactly three cases:

1. All roots are real and distinct: $x_1, x_2, x_3 \in \mathbb{R}$.
2. All roots are real, but at least two of them coincide: $x_1, x_2, x_3 \in \mathbb{R}$.
3. One root is real, $x_1 \in \mathbb{R}$, and two are imaginary complex conjugate roots, $x_3 = \bar{x}_2$, $x_2, x_3 \in \mathbb{C}$.

From the definition of the discriminant of the equation $a_3 x^3 + a_2 x^2 + a_1 x + a_0 = 0$ in each of the three cases, we get:

1. $D = a_3^4 (x_1 - x_2)^2 (x_1 - x_3)^2 (x_2 - x_3)^2 > 0$.
2. $D = a_3^4 (x_1 - x_2)^2 (x_1 - x_3)^2 (x_2 - x_3)^2 = 0$.
3. $D = a_3^4 (x_1 - x_2)^2 (x_1 - x_3)^2 (x_2 - x_3)^2 = a_3^4 \left((x_1 - x_2)(x_1 - \bar{x}_2) \right)^2 (x_2 - \bar{x}_2)^2 = -a_3^4 |x_1 - x_2|^4 (\Im x_2)^2 < 0$.

The result of the roots of the cubic equation is again given in the form of a table.

$D > 0$	$D = 0$	$D < 0$
Three various real roots	Roots real and at least two of them coincide	One real root and two complex-conjugate

Example 2.23. Does the polynomial have complex roots

$$x^3 + x^2 + x + 1 = 0?$$

Proof. According to formula (2.11) for the discriminant, we have

$$D = 1 - 4 - 4 + 18 - 27 = -16 < 0.$$

Therefore, the equation has one real and two complex conjugate roots. Note that this equation can be easily solved. In fact,

$$x^3 + x^2 + x + 1 = 0 \iff x^2(x+1) + (x+1) = 0$$
$$\iff (x+1)(x^2+1) = 0 \iff x_1 = -1, \quad x_{2,3} = \pm i. \qquad \square$$

2.8.2 Solutions of the fourth-degree equation (using Cardano formulas)

The equation of the fourth-degree equation $ax^4 + bx^3 + cx^2 + dx + e = 0$, $a \neq 0$, can be easily reduced to an equation of the form $x^4 + px^2 + qx + r = 0$. It is enough to first divide by a, and then make a substitution of $y = x + a_1/4$, where $a_1 = b/a$. For the equation $z^3 - pz^2 - 4rz + (4pro^2) = 0$, denote its roots as z_1, z_2, z_3. The roots z_1, z_2, z_3 of the equation $z^3 - pz^2 - 4rz + (4pro^2) = 0$ are associated with the roots x_1, x_2, x_3, x_4 of the equation $x^4 + px^2 + qx + r = 0$ through the relations

$$z_1 = x_1x_2 + x_3x_4, \quad z_2 = x_1x_3 + x_2x_4, \quad z_3 = x_1x_4 + x_2x_3.$$

The equation $z^3 - pz^2 - 4rz + (4pr - q^2) = 0$ is called the cubic resolvent of the equation $x^4 + px^2 + qx + r = 0$. The roots z_1, z_2, z_3 are found using the Cardano formulas. Then, proving the relations

$$(x_1 + x_2 - x_3 - x_4)^2 = 4(z_1 - p),$$
$$(x_1 - x_2 + x_3 - x_4)^2 = 4(z_2 - p),$$
$$(x_1 - x_2 - x_3 + x_4)^2 = 4(z_3 - p),$$
$$(x_1 + x_2 - x_3 - x_4)(x_1 - x_2 + x_3 - x_4)(x_1 - x_2 - x_3 + x_4) = -8q,$$

we get

$$x_1 + x_2 + x_3 + x_4 = 0,$$
$$x_1 + x_2 - x_3 - x_4 = 2\sqrt{z_1 - p},$$
$$x_1 - x_2 + x_3 - x_4 = 2\sqrt{z_2 - p},$$
$$x_1 - x_2 - x_3 + x_4 = 2\sqrt{z_3 - p},$$

from which we obtain formulas for finding the roots of the fourth-degree polynomial:

$$x_1 = \frac{1}{2}\left(\sqrt{z_1 - p} + \sqrt{z_2 - p} + \sqrt{z_3 - p}\right);$$

$$x_2 = \frac{1}{2}\left(\sqrt{z_1 - p} - \sqrt{z_2 - p} - \sqrt{z_3 - p}\right);$$

$$x_3 = \frac{1}{2}\left(-\sqrt{z_1 - p} - \sqrt{z_2 - p} + \sqrt{z_3 - p}\right);$$

$$x_4 = \frac{1}{2}\left(-\sqrt{z_1 - p} + \sqrt{z_2 - p} - \sqrt{z_3 - p}\right).$$

Moreover, we choose the values of the square roots in such a way that their product equals $-q$, i.e., the following equality holds

$$\sqrt{z_1 - p} \cdot \sqrt{z_2 - p} \cdot \sqrt{z_3 - p} = -q.$$

The roots z_1, z_2, z_3 of the equation $z^3 - pz^2 - 4rz + (4pr - q^2) = 0$ can be calculated, for example, using the Cardano formulas.

Example 2.24. Solve the equation

$$x^4 - 5x^2 + 4 = 0.$$

Proof. Let us make a cubic resolvent of this equation:

$$p = -5, \quad q = 0, \quad r = 4;$$

$$z^3 + 5z^2 - 16z - 80 = 0 \iff z^2(z + 5) - 16(z + 5) = 0$$

$$\iff (z^2 - 16)(z + 5) = 0.$$

Therefore,

$$z_1 = -5, \quad z_2 = -4, \quad z_3 = 4.$$

Let's choose the roots as follows[4]:

$$\sqrt{z_1 - p} = \sqrt{-5 + 5} = 0,$$

$$\sqrt{z_2 - p} = \sqrt{-4 + 5} = 1,$$

$$\sqrt{z_3 - p} = \sqrt{4 + 5} = 3.$$

---

[4]Do not forget that $\sqrt{z}$ always has two values for $z \neq 0$, $z \in \mathbb{C}$.

Indeed, the equality $\sqrt{z_1 - p} \cdot \sqrt{z_2 - p} \cdot \sqrt{z_3 - p} = -q = 0$ holds. Finally, we find

$$x_1 = \frac{1}{2}\left(\sqrt{z_1 - p} + \sqrt{z_2 - p} + \sqrt{z_3 - p}\right) = \frac{1}{2}(0 + 1 + 3) = 2;$$

$$x_2 = \frac{1}{2}\left(\sqrt{z_1 - p} - \sqrt{z_2 - p} - \sqrt{z_3 - p}\right) = \frac{1}{2}(0 - 1 - 3) = -2;$$

$$x_3 = \frac{1}{2}\left(-\sqrt{z_1 - p} - \sqrt{z_2 - p} + \sqrt{z_3 - p}\right) = \frac{1}{2}(-0 - 1 + 3) = 1;$$

$$x_4 = \frac{1}{2}\left(-\sqrt{z_1 - p} + \sqrt{z_2 - p} - \sqrt{z_3 - p}\right) = \frac{1}{2}(-0 + 1 - 3) = -1.$$

Thus, we have found the roots of the equation $x^4 - 5x^2 + 4 = 0$.

In this case, we could have arrived at the same result in a much easier way:

$$x^4 - 5x^2 + 4 = 0 \iff (x^2 - 1)(x^2 - 4) = 0$$
$$\iff (x - 1)(x + 1)(x - 2)(x + 2) = 0. \qquad \square$$

Example 2.25. Solve the equation

$$x^4 + 4x^3 + 9x^2 + 16x + 20 = 0.$$

Proof. Since the coefficient at x^3 is different from zero, you need to replace $y = x + b/4a = x + 1$. We get

$$x^4 + 4x^3 + 9x^2 + 16x + 20 = 0$$
$$\iff (y - 1)^4 + 4(y - 1)^3 + 9(y - 1)^2 + 16(y - 1) + 20 = 0$$
$$\iff (y - 4y^3 + 6y^2 - 4y + 1) + 4(y^3 - 3y^2 + 3y - 1)$$
$$+ 9(y^2 - 2y + 1) + 16y + 4 = 0$$
$$\iff y^4 + 3y^2 + 6y + 10 = 0.$$

Let us find the roots of the last equation. Let us make a cubic resolvent of this equation:

$$p = 3, \quad q = 6, \quad r = 10;$$
$$z^3 - 3z^2 - 40z + 84 = 0.$$

From this, according to the Cardano formulas, or from decomposition $z^3 - 3z^2 - 40z + 84 = (z + 6)(z - 2)(z - 7)$, we find

$$z_1 = -6, \quad z_2 = 2, \quad z_3 = 7.$$

Let's choose the roots as follows[5]:

$$\sqrt{z_1 - p} = \sqrt{-6 - 3} = 3i,$$
$$\sqrt{z_2 - p} = \sqrt{2 - 3} = i,$$
$$\sqrt{z_3 - p} = \sqrt{7 - 3} = 2.$$

Indeed, the equality $\sqrt{z_1 - p} \cdot \sqrt{z_2 - p} \cdot \sqrt{z_3 - p} = -q$ is true because $3i \cdot i \cdot 2 = -6$. Finally, we find

$$y_1 = \frac{1}{2}\left(\sqrt{z_1 - p} + \sqrt{z_2 - p} + \sqrt{z_3 - p}\right)$$
$$= \frac{1}{2}(3i + 3 + 2) = 1 + 2i;$$

$$y_2 = \frac{1}{2}\left(\sqrt{z_1 - p} - \sqrt{z_2 - p} - \sqrt{z_3 - p}\right)$$
$$= \frac{1}{2}(3i - i - 2) = -1 + i;$$

$$y_3 = \frac{1}{2}\left(-\sqrt{z_1 - p} - \sqrt{z_2 - p} + \sqrt{z_3 - p}\right)$$
$$= \frac{1}{2}(-3i - i + 2) = 1 - 2i;$$

$$y_4 = \frac{1}{2}\left(-\sqrt{z_1 - p} + \sqrt{z_2 - p} - \sqrt{z_3 - p}\right)$$
$$= \frac{1}{2}(-3i + i - 2) = -1 - i.$$

Thus, we found the roots of the equation $y^4 + 3y^2 + 6y + 10 = 0$. By replacing $y = x + 1$, we get $x = y - 1$. Therefore,

$$x_1 = 2i, \quad x_2 = -2 - i, \quad x_3 = -2i, \quad x_4 = -2 - i$$

are the roots of the original equation.

Indeed, it is easy to verify that $x_1 = 2i$, $x_2 = -2 - i$, $x_3 = -2i$, and $x_4 = -2 - i$ are the roots of the original equation. For example, it follows from the representation

$$x^4 + 4x^3 + 9x^2 + 16x + 20 = (x^2 + 4)(x^2 + 4x + 5). \qquad \Box$$

---

[5]Do not forget that $\sqrt{z}$ with $z \neq 0$, $z \in \mathbb{C}$ always has two values, so, in this case, $\sqrt{-6 - 3} = \pm 3i$, $\sqrt{2 - 3} = \pm i$, $\sqrt{7 - 3} = \pm 2$.

2.8.2.1   *Discriminant of the fourth-degree equation*

Let x_1, x_2, x_3, x_4 be the roots of the equation $a_4x^4 + a_3x^3 + a_2x^2 + a_1x + a_0 = 0$, $a_4 \neq 0$, $a_4, a_3, a_2, a_1, a_0 \in \mathbb{R}$. If a polynomial with real coefficients has a complex root, then the complex-conjugate number with it will necessarily be the root of the equation. Therefore, there are exactly three cases:

1. All roots are distinct and either real, $x_1, x_2, x_3, x_4 \in \mathbb{R}$, or complex x_1, $x_2, x_3, x_4 \in \mathbb{C}$. Moreover, in the case of complex numbers, we have two pairs of complex-conjugate roots, $x_2 = \bar{x}_1$, $x_4 = \bar{x}_3$, $x_1, x_2, x_3, x_4 \in \mathbb{C}$.
2. At least two roots coincide: $x_k = x_l$, $k \neq l$, $k, l = 1, 2, 3, 4$.
3. There are two real roots, $x_1, x_2 \in \mathbb{R}$, and two imaginary complex-conjugate roots, $x_4 = \bar{x}_3$, $x_3, x_4 \in \mathbb{C}$.

From the definition of the discriminant for the equation $a_3x^3 + a_2x^2 + a_1x + a_0 = 0$ in each of the three cases, we get the following:

1. For real roots, it is obvious that $D > 0$. For complex roots,

$$D = a_4^4(x_1 - x_2)^2(x_1 - x_3)^2(x_1 - x_4)^2$$
$$\cdot (x_2 - x_3)^2(x_2 - x_4)^2(x_3 - x_4)^2$$
$$= a_4^4 (i \cdot \Im x_1)^2 (i \cdot \Im x_3)^2 \cdot |x_1 - x_3|^4 |x_1 - \bar{x}_3|^4$$
$$= a_4^4 (i \cdot \Im x_1)^2 (i \cdot \Im x_3)^2 |x_1 - x_3|^4 \cdot |x_1 - \bar{x}_3|^4 > 0.$$

2. $D = a_4^4(x_1 - x_2)^2(x_1 - x_3)^2(x_1 - x_4)^2(x_2 - x_3)^2(x_2 - x_4)^2 \cdot (x_3 - x_4)^2 = 0.$
3. $D = a_4^4(x_1 - x_2)^2(x_1 - x_3)^2(x_1 - x_4)^2(x_2 - x_3)^2(x_2 - x_4)^2 \cdot (x_3 - x_4)^2 = a_4^4(x_1 - x_2)^2|x_1 - x_3|^4|x_2 - x_3|^4 (i \cdot \Im x_3)^2 = -a_4^4(x_1 - x_2)^2|x_1 - x_3|^4|x_2 - x_3|^4 (\Im x_3)^2 < 0.$ The result of the roots of the equation of the fourth degree is again given in the form of a table.

$D > 0$	$D = 0$	$D < 0$
The roots are different and either all real or there are two pairs of complex-conjugate roots	Among the roots there are at least two identical	Two real and two complex-conjugate roots

2.8.2.2   *Discriminant of an equation of arbitrary degree*

We have the following features:

1. All roots are distinct and either real or complex. Moreover, any complex root has a complex-conjugate root.
2. At least two roots coincide.

From the definition of the discriminant, as before, it is not difficult to prove[6] that the following representations are valid for the discriminant:

1. $\operatorname{sign} D = (-1)^k$, where k is the number of pairs of complex conjugate roots. If $k = 1$, then we have two roots that are complex conjugate and two real roots, and if $k = 2$, then we have two pairs of complex-conjugate roots (four complex roots in total).
2. $D = 0$.

The result of the roots of the equation of arbitrary degree is again given in the form of a table.

$\operatorname{sign} \mathbf{D} = (-\mathbf{1})^{\mathbf{k}}$	$\mathbf{D} = \mathbf{0}$
The roots are different and among the roots there are exactly k pairs complex-conjugate	Among the roots there are at least two of the same root

---

Problems

2.1. Solve the equation

$$2x^4 + x^3 - 11x^2 + x + 2 = 0.$$

2.2. Solve the equation

$$x^4 - 5x^3 + 10x^2 - 10x + 4 = 0.$$

2.3. Solve the equation

$$78x^4 - 133x^3 + 78x^2 - 133x + 78 = 0.$$

2.4. Solve the equation

$$(x + 5)^4 - 13x^2(x + 5)^2 + 36x^4 = 0.$$

2.5. Solve the equation

$$2(x^2 + x + 1)^2 - 7(x - 1)^2 = 13(x^3 - 1).$$

2.6. Solve the equation

$$x^3 + 5x^2 + 15x + 27 = 0.$$

2.7. Solve the equation

$$27x^3 - 15x^2 + 5x - 1 = 0.$$

2.8. Solve the equation

$$x^3 + 1991x + 1992 = 0.$$

2.9. Solve the equation

$$x^3 - x^2 - 81x + 81 = 0.$$

2.10. Solve the equation

$$2x + 1 + \frac{4x^4}{2x + 1} = 5x^2.$$

2.11. Solve the equation

$$\frac{x^2}{1 - 2x^2} = 12x^2 + 7x - 6.$$

2.12. Solve the equation

$$x^4 - x^3 - 10x^2 + 2x + 4 = 0.$$

2.13. Prove that the system of equations

$$\begin{cases} 3x^3 + 13x^2 + 20x + 14 = 0, \\ (6x + 17)^y - 5 = \dfrac{7y}{x} + 2^{x+y} \\ \qquad \cdot \sqrt{9x(x+3)^2 - 3x^2 + 10x + 49} \end{cases}$$

has a unique solution.

2.14. Find all the values of a for each of which, among the solutions of the equation

$$(a^4 + 2014a^3 + 2014a^2 + 2014a + 2013)x = a^3 + 3a^2 - 6a - 8,$$

there are non-negative numbers.

2.15. Find the largest value of a for which the equation $x^3 + 5x^2 + ax + b = 0$ with integer coefficients has three different roots, one of which is equal to -2.

2.16. For what values of a the inequality

$$(x^2 - (a + 8)x - 6a^2 + 24a)\sqrt{3 - x} \le 0$$

has a unique solution?

2.17. For each value of a, solve the inequality

$$(x^2 + 2x - a^2 - 4a - 3)(\sin x + 2x) > 0.$$

2.18. For each value of a, solve the inequality

$$\frac{x^2 \cdot 2^{|2a-1|} - 2x + 1}{x^2 - (a-2)x - 2a} > 0.$$

2.19. Find all the values of a for each of which there is exactly one solution to the inequality

$$x^3\sqrt{a^3 + a^2 - a - 1} - x^2\sqrt{a^3 + a^2} + x\sqrt{a^4 - a^2} - a^2 \le 0$$

satisfying the condition $a \le x \le 2a + 1$.

2.20. For each value of a, solve the equation

$$2x^2 + 3x - a + 4 = \sqrt{-x^3 + (a-1)x^2 + (a-2)x + 2a}.$$

2.21. For each value of a, solve the inequality

$$2ax^4 + 8x^3 + (a + 2a^3)x^2 + 4x + a^3 > 0.$$

2.22. Find all the values of β for which the equation

$$(x^2 + x)(x^2 + 5x + 6) = \beta$$

with respect to x has exactly three different roots.

2.23. Find all values of $y > 1/2$ such that the inequality

$$16y^3 + 6y^3x - 4y^3x^2 - 50y^2 - 11y^2x + 10y^2x^2 + 52y + 4yx$$
$$- 8yx^2 - 18 + x + 2x^2 > 0$$

is executed for all x on the interval $1 < x < 2y$.

To solve Problems 2.24–2.32, use Vieta's theorem.

2.24. The quadratic equation

$$x^2 - 6px + q = 0$$

has two distinct roots x_1 and x_2. The numbers p, x_1, x_2, and q are four consecutive terms of a geometric progression. Find x_1 and x_2.

2.25. For what values of a the four roots of the equation

$$x^4 + (a - 5)x^2 + (a + 2)^2 = 0$$

are consecutive members of an arithmetic progression?

2.26. One of the roots of the equation $ax^2 + bx + 2 = 0$, where $a < 0$, is equal to $x = 3$. Solve the equation

$$ax^4 + bx^2 + 2 = 0.$$

2.27. Find all the values of u and v for which there are two different roots of the equation

$$x(x^2 + x - 8) = u,$$

which are also the roots of the equation

$$x(x^2 - 6) = v.$$

2.28. Determine all the values of a for each of which there are three different roots of the equation

$$x^3 + (a^2 - 15a)x^2 + 12ax - 216 = 0$$

that form a geometric progression. Find these roots.

2.29. What values, depending on a, can the expression $x_1^2 + x_1 x_2 + x_2^2$ take, in which x_1 and x_2 are two different roots of the equation

$$x^3 - 2007x = a?$$

2.30. Find all the values of a for each of which the equation

$$x^3 - ax^2 - (a^3 - 6a^2 + 5a + 8)x - (a - 3)^3 = 0$$

has three different roots forming a geometric progression (specify these roots).

2.31. Three equations with real coefficients are given:

$$x^2 - (a + b)x + 8 = 0,$$
$$x^2 - b(b + 1)x + c = 0,$$
$$x^4 - b(b + 1)x^2 + c = 0.$$

Each of them has at least one real root. It is known that the roots of the first equation are greater than one. It is also known that all the roots of the first equation are the roots of the third equation and at least one root of the first equation satisfies the second equation. Find the numbers a, b, and c if $b > 3$.

2.32. Find the sum of the squares of all the real roots of the equation

$$x^5 + 2010x^2 + 2011 = x^4 + 2011x^3 + 2012x.$$

For Problems 2.33–2.52, find all solutions in terms of complex numbers $\mathbb{C}$.

2.33. Solve the equation $x^2 + 64 = 0.$

2.34. Solve the equation $x^2 + 6x + 10 = 0.$

2.35. Solve the equation $x^2 - 5x - 150 = 0.$

2.36. Solve the equation $x^2 + 14x + 58 = 0.$

2.37. Solve the equation $x^3 - 2x + 4 = 0.$

2.38. Solve the equation $x^3 - 3x + 52 = 0.$

2.39. Solve the equation $x^3 + 37x - 212 = 0$.

2.40. Solve the equation $x^3 + 9x^2 - 34x - 336 = 0$.

2.41. Solve the equation $x^3 + 4x^2 + 21x + 34 = 0$.

2.42. Solve the equation $x^3 + 11x^2 + 23x + 45 = 0$.

2.43. Solve the equation $x^3 + 2.92x - 5.232 = 0$.

2.44. Solve the equation $x^3 + 2\sqrt{3}x^2 - 12x - 24\sqrt{3} = 0$.

2.45. Solve the equation $x^3 - \sqrt{7}x^2 + 5x - 5\sqrt{7} = 0$.

2.46. Solve the equation $x^4 + 8x^2 - 16x + 20 = 0$.

2.47. Solve the equation $x^4 + 18x^2 - 24x + 85 = 0$.

2.48. Solve the equation $x^4 - 37x^2 - 74x - 70 = 0$.

2.49. Solve the equation $x^4 + 2x^3 + x^2 - 2x - 2 = 0$.

2.50. Solve the equation $x^4 - 2x^3 + 14x^2 - 18x + 45 = 0$.

2.51. Solve the equation $x^4 - 2x^3 - 62x^2 + 128x - 128 = 0$.

2.52. Solve the equation $x^4 - 14x^3 + 51x^2 - 28x + 98 = 0$.

Hints and Answers

2.1. $2, 1/2, (-3 \pm \sqrt{5})/2$.

2.2. $1, 2$.

2.3. $2/3, 3/2$.

2.4. $-5/3, -5/4, 5/2, 5$.

2.5. $-1, -1/2, 2, 4$.

2.6. -3.

2.7. $1/3$.

2.8. -1.

2.9. $\pm 9, 1$.

2.10. $(1 \pm \sqrt{5})/4, 1 \pm \sqrt{2}$.

2.11. $-1, -3/4, 1/2, 2/3$.

2.12. $-1 \pm \sqrt{3}, (3 \pm \sqrt{17})/2$.

2.14. $(-2013; -4] \cup \{-1\} \cup [2; +\infty)$.

2.15. $a = 7$.

2.16. $a \in [1; 5/2]$.

2.17. If $a \le -3$, then $x \in (a+1; 0) \cup (-(a+3); +\infty)$; if $a \in (-3; -2)$, then $x \in (a+1; -(a+3)) \cup (0; +\infty)$; if $a = -2$, then $x \in (0; +\infty)$; if $a \in (-2; -1)$, then $x \in (-(a+3); a+1) \cup (0; +\infty)$; if $a \ge -1$, then $x \in (-(a+3); 0) \cup (a+1; +\infty)$.

2.18. If $a < -2$, then $x \in (-\infty; a) \cup (-2; +\infty)$; if $a = -2$, then $x \in \mathbb{R} \setminus \{-2\}$; if $a \in (-2; 1/2)$, then $x \in (-\infty; -2) \cup (a; +\infty)$; if $a = 1/2$, then $x \in (-\infty; -2) \cup (1/2; 1) \cup (1; +\infty)$; if $a > 1/2$, then $x \in (-\infty; -2) \cup (a; +\infty)$.

2.19. $a = 1, a = \sqrt{2}$. **Hint:** Find the domain of definition for the inequality by a and solve it for the given values of a.

2.20. If $a < 1$, then there are no solutions; if $a = 1$, then there is only one solution $x = -1$; if $a > 1$, then solutions $x = -1 \pm \sqrt{a-1}$.

2.21. If $a \le -\sqrt{2}$, then there are no solutions; if $-\sqrt{2} < a < 0$, then $x \in ((-2 + \sqrt{4 - a^4})/a; (-2 - \sqrt{4 - a^4})/a)$; if $a = 0$, then $x \in (0; +\infty)$; if $0 < a \le \sqrt{2}$, then $x \in (-\infty; (-2 - \sqrt{4 - a^4})/a) \cup ((-2 + \sqrt{4 - a^4})/a; +\infty)$; if $a > \sqrt{2}$, then $x \in (-\infty; +\infty)$.

2.22. $\beta = 9/16$.

2.23. $y \in [5/6; 1) \cup (1; 3/2]$. **Hint:** write down the equation as a square (with respect to x).

2.24. $(x_1; x_2) = (-3; 9); (2; 4)$.

2.25. $a = -5$, $a = -5/13$.

2.26. $x = \pm\sqrt{3}$.

2.27. $u = 6$; $v = 4$.

2.28. $a = 13$; the roots of the equation 2; 6; 18.

2.29. 2007, if $|a| \le 2 \cdot (669)^{3/2}$.

2.30. If $a = 2$, then the roots of the equation $x = (3 - \sqrt{5})/2$; $x = -1$; $x = (3 + \sqrt{5})/2$; if $a = 4$, then $(3 - \sqrt{5})/2, 1, (3 + \sqrt{5})/2$.

2.31. $a = 2$, $b = 4$, $c = 64$.

2.32. 4025. **Hint:** Note that one of the multipliers of the fifth degree equation will be $x^2 + 1$.

2.33. $x_{1,2} = \pm 8i$.

2.34. $x_{1,2} = -3 \pm i$.

2.35. $x_1 = -10$, $x_2 = 15$.

2.36. $x_{1,2} = -7 \pm 3i$.

2.37. $x_1 = -2$, $x_{2,3} = 1 \pm 2i$.

2.38. $x_1 = -4$, $x_{2,3} = 2 \pm 3i$.

2.39. $x_1 = -4$, $x_{2,3} = -2 \pm 7i$.

2.40. $x_1 = -8$, $x_2 = -7$, $x_3 = 6$.

2.41. $x_1 = -2$, $x_{2,3} = -1 \pm 4i$.

2.42. $x_1 = -9$, $x_{2,3} = -1 \pm 2i$.

2.43. $x_1 = 6/5$, $x_{2,3} = -3/5 \pm 4i$.

2.44. $x_1 = 2\sqrt{3}$, $x_2 = x_3 = -2\sqrt{3}$.

2.45. $x_1 = \sqrt{7}$, $x_{2,3} = \pm\sqrt{5}i$.

2.46. $x_{1,2} = -1 \pm 3i$, $x_{3,4} = 1 \pm i$.

2.47. $x_{1,2} = 1 \pm 2i$, $x_{3,4} = -1 \pm 4i$.

2.48. $x_1 = -5$, $x_2 = 7$, $x_{3,4} = -1 \pm i$.

2.49. $x_1 = 1$, $x_2 = -1$, $x_{3,4} = -1 \pm i$.

2.50. $x_{1,2} = 1 \pm 2i$, $x_{3,4} = \pm 3i$.

2.51. $x_{1,2} = 1 \pm i$, $x_{3,4} = \pm 8$.

2.52. $x_1 = x_2 = 7$, $x_{3,4} = \pm\sqrt{2}i$.

Chapter 3

Equations and Inequalities with Modulus

The modulus (or absolute value) of a number a is the value

$$|a| = \begin{cases} a & \text{if } a > 0, \\ 0 & \text{if } a = 0, \\ -a & \text{if } a < 0. \end{cases} \qquad (3.1)$$

Obviously, $|-a| = |a|$ and $|a| \geq 0$, for any a. From this definition, it immediately follows that $|b - a|$ is the distance between the points a and b on the real axis.

We need the following facts:

$$|ab| = |a| \cdot |b|, \quad |a + b| \leq |a| + |b|, \quad |a - b| \leq |a| + |b|. \qquad (3.2)$$

These statements immediately follow from the above definition. Note that $|a + b| = |a| + |b|$ if the numbers a and b are both non-negative or non-positive, and $|a + b| < |a| + |b|$ if the numbers a and b have different signs.

Similarly, $|a - b| < |a| + |b|$ if a and b are of the same sign, and $|a - b| = |a| + |b|$ otherwise.

The graph of the function $y = |x|$ has the form shown in Fig. 3.1. This function is even, i.e., $|-x| = |x|$, and it is continuous everywhere. It has no derivative at the point $x = 0$.

For $x > 0$, the function $y = |x|$ coincides with the function $y = x$, and its derivative is 1; similarly, we obtain that for $x < 0$, the derivative of the function $|x|$ is -1.

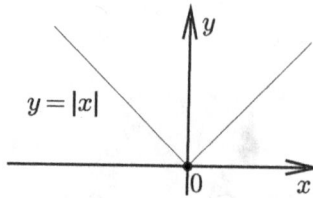

Fig. 3.1.   Graph of modulus.

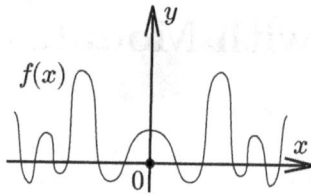

Fig. 3.2.   Graph of $f(x)$.

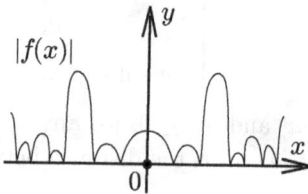

Fig. 3.3.   Graph of $|f(x)|$.

A more general situation involves obtaining the graph of the function $y = |f(x)|$ from the graph of $y = f(x)$ (Figs. 3.2 and 3.3). To do so, we replace all negative values of the function $f(x_0)$ by $-f(x_0)$, i.e., the negative values are symmetrically reflected relative to the straight line $y = 0$.

3.1   Interval Method for Equations and Inequalities with Modulus

Example 3.1. Solve the equation

$$|x| = x + 5.$$

Proof. For a solution, let us consider two cases. First, let $x < 0$. Then, according to (3.1), $|x| = -x$, and we get the equation

$$-x = x + 5, \quad \text{or} \quad 2x = -5,$$

from which we get $x = -5/2$. This number is located within the range under consideration, $x < 0$, and is part of the answer. Second, if $x \geq 0$, then $|x| = x$, and the equation takes the form

$$x = x + 5, \quad \text{or} \quad 0 = 5.$$

An incorrect equality is obtained, so the equation has no solutions in the domain under consideration, $x \geq 0$. As a result, we get $x = -5/2$. $\quad\square$

Example 3.2. Solve the equation

$$|x - 1| + |x - 2| = 1.$$

Proof. Let us begin by considering the fact that

$$|x - 1| = \begin{cases} x - 1 & \text{if } x \geq 1, \\ -x + 1 & \text{if } x < 1; \end{cases}$$

$$|x - 2| = \begin{cases} x - 2 & \text{if } x \geq 2, \\ -x + 2 & \text{if } x < 2. \end{cases}$$

Therefore, it is natural to consider three intervals on the axis (see Fig. 3.4):

1. Let $x < 1$. Then, $|x - 1| = 1 - x$, and $|x - 2| = 2 - x$. Substituting the obtained values into the original equation, we get $1 - x + 2 - x = 1$, or $x = 1$. The specified interval does not include this root.
2. Let $1 \leq x \leq 2$. Then, $|x - 1| = x - 1$, and $|x - 2| = 2 - x$. Substituting the obtained values into the original equation, we get $x - 1 + 2 - x = 1$, or $0 = 0$. The entire segment $[1; 2]$ will be part of the answer.

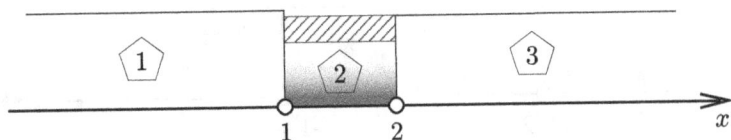

Fig. 3.4. The interval under consideration.

3. Let $x > 2$. Then, $|x - 1| = x - 1$, and $|x - 2| = x - 2$. Substituting the obtained values into the original equation, we get $x - 1 + x - 2 = 1$, or $x = 2$. The specified interval does not include this root.

Answer: $[1; 2]$.   □

Example 3.3. Solve the equation

$$|x^2 + 2x - 1| = |x + 1|.$$

Proof. Let us apply the interval method. Solving the equation

$$x^2 + 2x - 1 = 0,$$

its roots are $x_1 = -1 + \sqrt{2}$ and $x_2 = -1 - \sqrt{2}$. The second equation,

$$x + 1 = 0,$$

has the root $x_3 = -1$. Let's plot the numbers $-1 - \sqrt{2}$ and $-1, -1 + \sqrt{2}$ on the real axis and consider four intervals (see Fig. 3.5):

1. Let $x < -1 - \sqrt{2}$. Then, $x^2 + 2x - 1 > 0$ and $x + 1 < 0$. Therefore, $|x^2 + 2x - 1| = x^2 + 2x - 1$, $|x + 1| = -1 - x$, and we get the equation $x^2 + 2x - 1 = -1 - x$, or $x^2 + 3x = 0$. The roots of this equation are $x_4 = -3$ and $x_5 = 0$. The specified interval includes the root $x_4 = -3$.
2. Let $-1 - \sqrt{2} \le x < -1$. Then, $x^2 + 2x - 1 \le 0$ and $x + 1 < 0$, hence $|x^2 + 2x - 1| = -x^2 - 2x + 1$, $|x + 1| = -1 - x$, and we have the equation $-x^2 - 2x + 1 = -1 - x$, or $x^2 + x - 2 = 0$. Its roots are $x_6 = -2$ and $x_7 = 1$. The condition $-1 - \sqrt{2} \le x < -1$ is satisfied by $x_6 = -2$.
3. Let $-1 \le x < -1 + \sqrt{2}$. In this case, $x^2 + 2x - 1 < 0$ and $x + 1 \ge 0$. Thus, $|x^2 + 2x - 1| = -x^2 - 2x + 1$, $|x + 1| = x + 1$, and the equation takes the form $-x^2 - 2x + 1 = x + 1$, or $x^2 + 3x = 0$, which gives $x_8 = -3$ and $x_9 = 0$. The interval under consideration includes $x_9 = 0$.

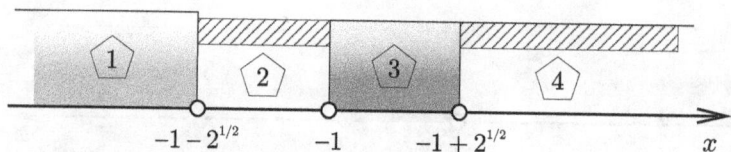

Fig. 3.5.   Interval method.

distance

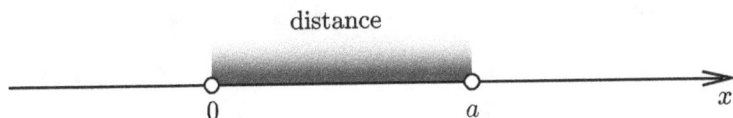

Fig. 3.6.   To Remark 3.1.

4. Let $x \geq -1 + \sqrt{2}$. Then, $x^2 + 2x - 1 \geq 0$ and $x + 1 > 0$. Therefore, $|x^2 + 2x - 1| = x^2 + 2x - 1$, $|x + 1| = x + 1$, and we get the equation $x^2 + 2x - 1 = x + 1$, or $x^2 + x - 2 = 0$. Its roots are $x_{10} = 1$ and $x_{11} = -2$. Of these, only $x_{10} = 1$ satisfies the condition $x \geq -1 + \sqrt{2}$.

Answer: $x = -3$, $x = -2$, $x = 0$, and $x = 1$. ☐

Remark 3.1. The solution to this problem can be obtained in another way, for which we note that the modulus of the number a is the distance on the numeric axis from the point with the coordinate a to the origin (see Fig. 3.6). Therefore, the equality $|b| = |a|$ holds if and only if $b = a$ or $b = -a$ since only the points a and $-a$ are located at a distance of $|a|$ from the origin. Therefore, instead of the original equation, it is enough to solve the following two equations: $x^2 + 2x - 1 = x + 1$, $x^2 + 2x - 1 = -x - 1$. After converting the first of the two, we get the equation $x^2 + x - 2 = 0$, with the roots $x_1 = -2$ and $x_2 = 1$. Solving the second equation, $x^2 + 3x = 0$, and we get $x_3 = 0$ and $x_4 = -3$.

The given example shows that geometric considerations sometimes allow us to obtain a shorter solution to a problem compared to the general interval method. Let us consider some examples using the fact that $|a - b|$ is the distance between the points with coordinates a and b on the real axis c.

Example 3.4. Solve the equation

$$|x - 5| = 3.$$

Proof. According to the above, we find points on the numerical axis, and the distance from them to the point with the coordinate 5 is 3 (see Fig. 3.7). These points have coordinates 2 and 8. ☐

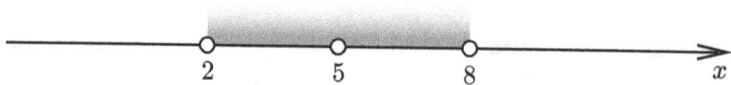

Fig. 3.7.   To example 3.4.

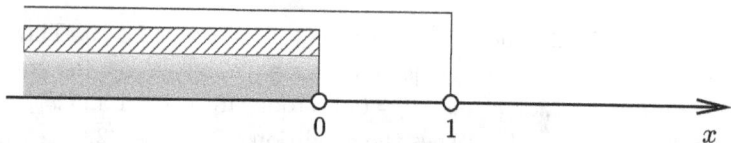

Fig. 3.8.   To example 3.5.

Example 3.5. Solve the inequality

$$|x - 1| > x + 1.$$

Proof. Consider the following cases:

- In the first case, $x < 1$. Then, the inequality takes the form $-x+1 > x+1$, or $0 > 2x$, or $x < 0$. We intersect the resulting set with the range under consideration (see Fig. 3.8), and we get part of the answer as $(-\infty; 0)$.
- In the second case, when $x \geq 1$, we get

$$x - 1 > x + 1 \Longleftrightarrow 0 > 2,$$

which is an incorrect inequality that has no solutions.

Answer: $x \in (-\infty; 0)$.   □

Example 3.6. Solve the inequality

$$|x - 2| + |x - 3| \leq 1.$$

Proof. We consider three intervals on the axis (see Fig. 3.9):

1. If $x < 2$, then the inequality will take the form

$$-x + 2 - x + 3 \leq 1 \Longleftrightarrow 4 - 2x \leq 0 \Longleftrightarrow x \geq 2.$$

The set of numbers $x \geq 2$ does not intersect with the domain under consideration, $x < 2$.

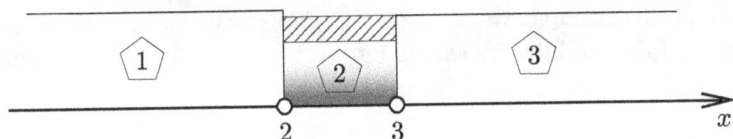

Fig. 3.9. Interval method.

2. If $2 \leq x \leq 3$, then the inequality will take the form

$$x - 2 - x + 3 \leq 1, \quad \text{or} \quad 1 \leq 1.$$

This inequality holds in the entire domain under consideration, $2 \leq x \leq 3$.

3. Finally, for $x > 3$, we get

$$x - 2 + x - 3 \leq 1, \quad \text{or} \quad 2x \leq 6, \quad \text{or} \quad x \leq 3.$$

Again, the resulting set does not intersect with the domain under consideration.

Answer: $x \in [2; 3]$. $\qquad\qquad\qquad\qquad\qquad\qquad\qquad\qquad$ □

Example 3.7. Solve the equality

$$x^2 + 1 + |x - 1| = 2|x|.$$

Proof. We transform this equation into the following form:

$$(|x|^2 - 2|x| + 1) + |x - 1| = 0,$$

or, equivalently, $(|x| - 1)^2 + |x - 1| = 0$. The sum of two non-negative expressions is zero if and only if each of them is zero:

$$\begin{cases} |x| = 1, \\ |x - 1| = 0. \end{cases} \iff x = 1.$$

$\qquad\qquad\qquad\qquad\qquad\qquad\qquad\qquad\qquad\qquad\qquad\qquad\qquad$ □

Example 3.8. Solve the equality

$$|x| = 2 - x.$$

Proof. In this example, you can split the entire real axis into two intervals, i.e., expand the modulus by taking into account the sign of the number x. Conversely, you can simply solve the equation

$$x = \pm(2 - x)$$

and then check by substitution whether we really found the roots of the equation.

Solving the equation $x = \pm(2-x)$, we find $x = 1$. Now, by substitution, we can easily confirm that we have found a solution to the equation.   □

Example 3.9. Solve the inequality

$$|x + |1 - x|| > 3.$$

Proof. Let us consider two cases depending on the sign of $1 - x$:

1. Let $1 - x \geq 0$. Then, the inequality takes the form

$$|x + 1 - x| > 3 \Longleftrightarrow 1 > 3.$$

   An incorrect inequality is obtained. There are no solutions in the range of $x \leq 1$.

2. In the case of $1 - x < 0$, i.e., $x > 1$, we get

$$|x - 1 + x| > 3 \Longleftrightarrow |2x - 1| > 3 \Longleftrightarrow 2x - 1 > 3 \Longleftrightarrow x > 2.$$

   Answer: $(2; +\infty)$.   □

Above, in Example 3.4, we used geometric considerations. To solve an equation of the form

$$|A(x)| = |B(x)|,$$

we replace this equation with an equivalent set of equations:

$$A(x) = B(x), \quad A(x) = -B(x).$$

Consider the inequality

$$|A(x)| < |B(x)|. \tag{3.3}$$

By the definition of the modulus, inequality (3.3) is equivalent to the inequality

$$A^2(x) < B^2(x) \iff A^2(x) - B^2(x) < 0$$
$$\iff (A(x) - B(x))(A(x) + B(x)) < 0. \tag{3.4}$$

Example 3.10. Solve the inequality

$$\left| \frac{x^2 - 3x - 1}{x^2 + x + 1} \right| < 3.$$

Proof. The application of the interval method will require us to consider intervals with boundaries $\frac{3 \pm \sqrt{13}}{2}$ and the intersection of the intervals obtained by solving for these intervals. However, using formulas (3.3) and (3.4), we transform our inequality into the inequality

$$\left(\frac{x^2 - 3x - 1}{x^2 + x + 1} - 3 \right) \left(\frac{x^2 - 3x - 1}{x^2 + x + 1} + 3 \right) < 0,$$

or

$$\frac{(-2x^2 - 6x - 4)(4x^2 + 2)}{(x^2 + x + 1)^2} < 0,$$

or

$$\frac{(x + 1)(x + 2)(4x^2 + 2)}{(x^2 + x + 1)^2} > 0,$$

and this inequality has solutions $x \in (-\infty; -2) \cup (-1; +\infty)$ since $4x^2 + 2 > 0$, for all x, and $x^2 + x + 1 > 0$, for all x. $\qquad\square$

Example 3.11. Solve the inequality

$$|x^3 - 6x^2 + 2x - 11| < |x^3 - 4x^2 + 10x + 11|.$$

Proof. An attempt to apply the interval method to this problem will encounter significant difficulties. Therefore, we use inequalities (3.3) and

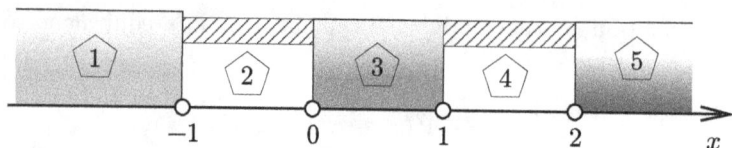

Fig. 3.10.   Interval method.

(3.4), and we find that the inequality in question is equivalent to the inequality

$$(-2x^2 - 8x - 22)(2x^3 - 10x^2 + 12x) < 0,$$

or

$$(x^2 + 4x + 11) \cdot x \cdot (x^2 - 5x + 6) > 0.$$

Since $x^2 + 4x + 11 > 0$, for all x, the original inequality is equivalent to the inequality

$$x(x - 2)(x - 3) > 0,$$

whose solutions are $x \in (0; 2) \cup (3; +\infty)$.   □

Let us consider an example in which an expression containing a modulus is placed within another modulus.

Example 3.12. Solve the equation

$$1 + ||x - 1| - 1| = 2|x + 1|.$$

Proof. We use the interval method. First, let's solve the equations

$$x - 1 = 0, \quad x + 1 = 0, \quad |x - 1| - 1 = 0.$$

The first two equations have solutions $x = 1$ and $x = -1$, respectively. The last one is similar to the equation solved in Example 3.4. Its solutions are points with coordinates 0 and 2, each separated from the point with coordinate 1 by a distance of 1.

Let us plot the points with coordinates -1, 0, 1, and 2 on the real axis and consider the following five intervals (see Fig. 3.10):

1. Let $x < -1$. Then,
$$|x+1| = -x-1 \quad \text{and} \quad |x-1| = 1-x,$$
$$||x-1|-1| = |1-x-1| = |-x| = -x,$$
and we get the equation
$$1-x = -2x-2.$$
Therefore, $x_1 = -3$.

2. Let $-1 \le x < 0$. Then,
$$|x+1| = x+1 \quad \text{and} \quad |x-1| = 1-x,$$
$$||x-1|-1| = |1-x-1| = |-x| = -x,$$
and the equation takes the form $1-x = 2x+2$, or $3x = -1$, which gives $x_2 = -1/3$.

3. Let $0 \le x < 1$. Then,
$$|x+1| = x+1 \quad \text{and} \quad |x-1| = 1-x,$$
$$||x-1|-1| = |1-x-1| = |-x| = x.$$
Therefore, we get the equation $1+x = 2x+2$, or $x = -1$. This number is not included in the segment under consideration.

4. Let $1 \le x < 2$. Then,
$$|x+1| = x+1 \quad \text{and} \quad |x-1| = x-1,$$
$$||x-1|-1| = |x-1-1| = |x-2| = 2-x,$$
and the original equation reduces to the equation
$$1+2-x = 2x+2 \Longleftrightarrow 3x = 1.$$
The solution $x = 1/3$ does not satisfy the condition $1 \le x < 2$.

5. Let $x \ge 2$. Then,
$$|x+1| = x+1 \quad \text{and} \quad |x-1| = x-1,$$
$$||x-1|-1| = |x-1-1| = |x-2| = x-2,$$
and
$$1+x-2 = 2x+2,$$
from which we get $x = -3$. The resulting value is not included in the interval $x \ge 2$.

Answer: $x_1 = -3$, $x_2 = -1/3$. $\qquad\qquad$ □

Example 3.13. Solve the equation

$$||||x - 1| + 2| - 1| + 1| = 2.$$

Proof. In this example, it is convenient to note that $||||x-1|+2|-1|+1| = |||x-1|+2|-1|+1$ since the modulus is non-negative. Therefore, the equation is equivalent to the equation $|||x - 1| + 2| - 1| = 1$.

As noted above, this equation is equivalent to a set of equations:

$$\left[\begin{array}{l} ||x - 1| + 2| - 1 = 1, \\ ||x - 1| + 2| - 1 = -1. \end{array}\right.$$

The first of them gives the equation $||x - 1| + 2| = 2$, which is similar to the one considered above and is equivalent to the equation

$$|x - 1| + 2 = 2, \text{ or } |x - 1| = 0, \text{ or } x = 1.$$

The equation $||x - 1| + 2| - 1 = -1$ gives the equation $||x - 1| + 2| = 0$, or $|x - 1| = -2$, which has no solutions. $\quad\square$

3.1.1 *Using the properties of modulus function $|x|$*

Note that for all A and B, the following inequalities hold:

$$|A| + A \geq 0, \quad |B| - B \geq 0. \tag{3.5}$$

Indeed, by the definition of the modulus,

$$|A| + A = \begin{cases} A + A = 2A > 0 & \text{if } A > 0, \\ -A + A = 0 & \text{if } A \leq 0; \end{cases}$$

$$|B| - B = \begin{cases} B - B = 0 & \text{if } B \geq 0, \\ -B - B = -2B > 0 & \text{if } B < 0. \end{cases} \tag{3.6}$$

Inequalities (3.5) and equalities (3.6) are useful in solving some problems.

Example 3.14. Let N be a natural number. Solve the equation

$$|x + 1| + |x - 1| + \cdots + |x + N| + |x - N| = 2Nx.$$

Proof. Since $2Nx = (x+1)+(x-1)+\cdots+(x+N)+(x-N)$, the original equation can be rewritten as

$$|x+1|+|x-1|+\cdots+|x+N|+|x-N| = (x+1)+(x-1)+\cdots+(x+N)+(x-N),$$

or

$$(|x+1| - (x+1)) + (|x-1| - (x-1))$$
$$+\cdots+ (|x+N| - (x+N)) + (|x-N| - (x-N)) = 0.$$

All the terms have the form $(|x+a| - (x+a))$ or $(|x-a| - (x-a))$ and are non-negative, according to (3.5). The sum of non-negative terms is 0 if and only if each of them is 0. According to (3.6), this means that all of the following inequalities are satisfied:

$$\begin{cases} x+1 \ge 0, \\ x-1 \ge 0, \\ \vdots \\ x+N \ge 0, \\ x-N \ge 0, \end{cases}$$

which means that the solutions of the equation are all $x \ge N$. $\qquad\square$

Example 3.15. For all a, solve the equation

$$|x - a^3 + 14a^2 - 3a + 5| + |x - a^3 + 14a^2 - 5a + 3| = 2a + 2.$$

Proof. Denote

$$A(x) = x - a^3 + 14a^2 - 3a + 5, B(x) = x - a^3 + 14a^2 - 5a + 3.$$

Of course, we will not solve the equation using the interval method; instead, we note that

$$A(x) - B(x) = 2a + 2.$$

This will bring the equation to the form

$$|A(x)| + |B(x)| = A(x) - B(x),$$

or

$$|A(x)| - A(x) + |B(x)| + B(x) = 0.$$

According to inequalities (3.5), $|A(x)| - A(x) \ge 0$ and $|B(x)| + B(x) \ge 0$, for all x. Further, in view of (3.6), this means that $A(x) \ge 0$ and $B(x) \le 0$, respectively. Thus, the set of solutions of the original equation is

$$a^3 - 14a^2 + 3a - 5 \le x \le a^3 - 14a^2 + 5a - 3,$$

and it is not empty if the left boundary does not exceed the right boundary, which means that $2a + 2 \ge 0$, or $a \ge -1$.

Answer: for $a \ge -1$, the answer is $x \in [a^3 - 14a^2 + 3a - 5; a^3 - 14a^2 + 5a - 3]$; otherwise, $a < -1$, and there are no solutions. $\square$

Example 3.16. Solve the equation

$$|x^3 + 7x^2 - 11x - 6| + |x^3 - 12x^2 - 5x + 3| = 18x^2 - 2x - 13.$$

Proof. In this equation, we denote

$$A(x) = x^3 + 7x^2 - 11x - 6, B(x) = x^3 - 12x^2 - 5x + 3.$$

At the same time,

$$A(x) - B(x) = 19x^2 - 6x - 9.$$

But

$$18x^2 - 2x - 13 = 19x^2 - 6x - 9 - (x^2 - 4x + 4)$$
$$= 19x^2 - 6x - 9 - (x-2)^2 = A(x) - B(x) - (x-2)^2.$$

As a result, the original equation takes the form

$$|A(x)| + |B(x)| - A(x) + B(x) + (x-2)^2 = 0,$$
$$(|A(x)| - A(x)) + (|B(x)| + B(x)) + (x-2)^2 = 0.$$

The left part consists of non-negative terms, and their sum is 0. So, each of them is 0, which is possible only if

$$x = 2, \quad |A(2)| - A(2) = 0, \quad |B(2)| + B(2) = 0,$$

that is, $A(2) \ge 0, B(2) \le 0$. Substituting $x = 2$ into the definitions of $A(x)$ and $B(x)$ provides justification for these inequalities.

Answer: $x = 2$. $\square$

Example 3.17. Find all values at which the equation

$$x^2 + |2x + a| + |2x - a| - a^2 = 0$$

has an odd number of solutions. Find these solutions.

Proof. The solution to this problem is based on the simple consideration that if the equation $f(x) = 0$, in which $f(x)$ is an even function, has an odd number of roots, then there must be a root $x = 0$ among its roots. Indeed, if x is a root of the equation $f(x) = 0$, then $-x$ is also a root of the equation since $f(-x) = f(x) = 0$.

The function $x^2 + |2x + a| + |2x - a| - a^2$ is even because $(-x)^2 + |-2x + a| + |-2x - a| - a^2 = x^2 + |2x - a| + |2x + a| - a^2$. We use the equality $|-A| = |A|$.

Substituting $x = 0$ into the equation under consideration, we obtain

$$|a| + |-a| - a^2 = 0 \Longleftrightarrow a^2 = 2|a|,$$

from which we find $a = 0$, $a = 2$, and $a = -2$.

- When $a = 0$, we get

$$x^2 + |2x| + |2x| = 0 \Longleftrightarrow$$
$$x^2 + 4|x| = 0 \Longleftrightarrow |x|(|x| + 4) = 0 \Longleftrightarrow x = 0.$$

- For $a = 2$, we get

$$x^2 + |2x + 2| + |2x - 2| - 4 = 0.$$

This equation is solved using the interval method, considering the cases $x < -1$, $-1 \leq x < 1$, and $x \geq 1$.

In the first case, we obtain the equation

$$x^2 - 2x - 2 - 2x + 2 - 4 = 0 \Longleftrightarrow x^2 - 4x - 4 = 0,$$

and $x = 2 \pm 2\sqrt{2}$, but none of these roots are included in the range $x < -1$.

In the second case, we get

$$x^2 + 2x + 2 - 2x + 2 - 4 = 0 \Longleftrightarrow x^2 = 0 \Longleftrightarrow x = 0.$$

If $x \geq 1$, then the equation takes the form

$$x^2 + 2x + 2 + 2x - 2 - 4 = 0,$$
$$x^2 + 4x - 4 = 0 \Longleftrightarrow x = -2 \pm 2\sqrt{2},$$

but these roots are not included in the range $x \geq 1$.

- For $a = -2$, the equation takes the form $x^2 + |2x - 2| + |2x + 2| - 4 = 0$, that is, it coincides with the above.

Answer: $a = 0$, $a = 2$, and $a = -2$, with all these values giving $x = 0$. □

Let us consider another type of problem based on the idea that an expression of the form $y = |A_1x + B_1| + \cdots + |A_nx + B_n|$ represents a function that, at the corresponding intervals, behaves as the linear function $y = Ax + B$, where A is equal to one of the following set of numbers:

$$\pm A_1 \pm \cdots \pm A_n.$$

Example 3.18. Find all values of a for which the inequality

$$||x + a^2| + ||x + 2a| + |2x + 3a||| \le 6 - 5x \qquad (3.7)$$

holds, for all $x \in [-6; 0]$.

Proof. According to the above, the function on the left side is a linear function on each of some set of intervals, and its angular coefficient on each of these intervals is one of the sets of numbers

$$\pm 1 \pm 1 \pm 2,$$

that is, not less than -4.

Therefore, the function $||x+a^2|+||x+2a|+|2x+3a|||+5x-6$ increases on the segment $[-6; 0]$, and the condition (3.7) for any $x \in [-6; 0]$, is equivalent to the validity of this condition only for $x = 0$, that is, to the inequality

$$||a^2| + ||2a| + |3a||| \le 6,$$

so

$$a^2 + |5a| \le 6 \iff (|a| - 1)(|a| + 6) \le 0$$
$$\iff |a| \le 1 \iff -1 \le a \le 1. \qquad \square$$

Problems

3.1. Solve the equation $|2x + 3| = x^2$.

3.2. Solve the inequality $|x^2 + 3x| + x^2 - 2 \geq 0$.

3.3. Solve the equation $|2x - 15| = 22 - |2x + 7|$.

3.4. Solve the equation $|2x + 9| - |x - 6| = 15$.

3.5. Solve the inequality $3|x + 2| - 4|x + 1| \leq 2$.

3.6. Solve the equation $|4x - |x - 2| + 3| = 16$.

3.7. Solve the inequality $|2x + 8| \geq 8 - |1 - x|$.

3.8. Solve the equation $\dfrac{|2x - 1|}{|x - 1|} = \dfrac{|2x + 1|}{|x + 1|}$.

3.9. Solve the inequality

$$\frac{1}{|7 - \log_3 3x|} + \frac{1}{|4 - \log_9 9x^2|} \leq \frac{1}{|\log_9 81x|}.$$

3.10. Solve the inequality

$$|x^2 + x - 2| + |x + 4| \leq x^2 + 2x + 6.$$

3.11. Solve the equation

$$\left| x^3 - x^2 + \frac{3}{2} \cdot x + 1 \right| = \left| x^3 + x^2 - \frac{3}{2} \cdot x - 1 \right|.$$

3.12. Solve the inequality $\dfrac{1}{|x - 2|} > \dfrac{1}{|x + 2|}$.

3.13. Solve the equation

$$20 \cdot 2^{2x + \frac{1}{2x}} - 21 \cdot 2^{x + \frac{1}{4x}} + 5 = 5 \cdot |3 \cdot 2^{x + \frac{1}{4x}} - 1|.$$

3.14. Solve the inequality

$$30^x (8 \cdot 5^x + 6^x) \leq |6^{3x} - 2^{2x+2} \cdot 3^{2x+1} \cdot 5^x + 4 \cdot 5^{x+1} \cdot 30^x|.$$

3.15. Solve the inequality $\left(\dfrac{1}{3}\right)^{|x|} < \left(\dfrac{1}{9}\right)^{|x+2|}$.

3.16. Solve the inequality $\dfrac{4}{|x + 1| - 2} \geq |x - 1|$.

3.17. Solve the inequality

$$\frac{|x - 4| - |x - 1|}{|x - 3| - |x - 2|} < \frac{|x - 3| + |x - 2|}{|x - 4|}.$$

3.18. Solve the equation

$$|x - 1| + |x + 1| + |x - 2| + |x + 2| + \cdots +$$
$$+ |x - 100| + |x + 100| + 200x = 0.$$

3.19. For each a, solve the equation $x|x + 1| + a = 0$.

3.20. For each a, solve the inequality $|x + 2a| \leq 1/x$.

3.21. For what values of a does the equation

$$2|x - a| + a - 4 + x = 0$$

have solutions, and do all the solutions of the equation satisfy the inequality $0 \leq x \leq 4$?

3.22. For each a, solve the equation $|x + 2| + a|x - 4| = 6$.

3.23. Find all the values of a for which the equation

$$|1 - ax| = 1 + (1 - 2a)x + ax^2$$

has exactly one solution.

3.24. Find all the values of a for which the inequality

$$x^2 + 4x + 6a|x + 2| + 9a^2 \leq 0$$

has at most one solution.

3.25. Find all the values of k for which the following is true:

$$2x - |x - k^2| = 11k - 3 \cdot |x + 4k|$$

(1) has no solutions, (2) has a finite nonempty set of solutions.

3.26. Determine for what values of a the equation

$$x - \frac{a}{2} = 4|4|x| - a^2|$$

has exactly three distinct roots. Find these roots.

3.27. For what values of a does the equation

$$2|x - 9a| - 2a^2 + 35 + x = 0$$

have no solutions? For what values of a does this equation have at least one solution, and do all solutions of this equation belong to the segment $[-30; 63]$?

3.28. Find all pairs of $(a; b)$ for which the equation

$$|x - \sin^2 a| + |x + \cos^2 4a - 2\sin a \cos \cos^4 4a| = b\left(a + \frac{3}{2}\pi\right)$$

has a unique solution.

3.29. Find all the values of a for each of which the inequality

$$\frac{1}{2}|a - 2| \cdot |x + a - 4| + \left(\frac{a^2 - 4a + 3}{|a - 2|} - |a - 2|\right) \cdot |x - 2|$$
$$+ \frac{1}{2}|a - 2| \cdot |x - a| \leq 1$$

is satisfied for exactly two different values of x.

3.30. Find all values of a for which the inequality

$$|x + a^2| + |x + 8a| + |5x + 7a| \leq 56 - 8x$$

holds, for all $x \in [-10; 0]$.

Hints and Answers

3.1. -1, 3.

3.2. $x \le -\dfrac{2}{3}$, $x \ge \dfrac{1}{2}$.

3.3. $[-7/2; 15/2]$.

3.4. -30; 4.

3.5. $[-8/7; 0]$.

3.6. $-17/5$, $11/3$.

3.7. $(-\infty; -5] \cup [-1; +\infty)$.

3.8. 0, $\pm 1/\sqrt{2}$.

3.9. $0 < x \le 1$, $x \ne 3^{-4}$.

3.10. $[-6; -1] \cup [0; +\infty]$.

3.11. 0, 2, $1/2$.

3.12. $(0; 2) \cup (2; +\infty)$.

3.13. $x_{1,2,3,4} = (A_{1,2} \pm \sqrt{A_{1,2}^2 - 1})/2$, where $A_1 = \log_2((9 - \sqrt{31})/10)$, $A_2 = \log_2(3/10)$.

3.14. $x \in (-\infty; 0] \cup [\log_{6/5} 4; \log_{6/5} 7] \cup [\log_{6/5} 12; +\infty)$.

3.15. $(-4; -4/3)$.

3.16. $(1; 3] \cup [-1 - \sqrt{8}; -3)$.

3.17. $(3; 4) \cup (4; 7)$.

3.18. $(-\infty; -100]$.

3.19. If $a < 0$, $x = (-1 + \sqrt{1 - 4a})/2$; if $a = 0$, $x = 0$; $x = -1$, if $a \in (0; 1/4)$, $x = (-1 - \sqrt{1 + 4a})/2$, $x = (-1 \pm \sqrt{1 - 4a})/2$; if $a = 1/4$, $x = (-1 - \sqrt{2})/2$, $x = -1/2$; if $a > 1/4$, $x = (-1 - \sqrt{1 + 4a})/2$.

3.20. If $a < -1$, then $x \in (0; -a - \sqrt{a^2 - 1}] \cup [-a + \sqrt{a^2 - 1}; -a + \sqrt{a^2 + 1}]$; if $a \ge -1$, then $x \in (0; -a + \sqrt{a^2 + 1}]$.

3.21. $a \in [4/3; 2]$.

3.22. If $a < -1$, $x = 4$; if $a = -1$, $x \ge 4$; if $a \in (-1; 1)$, $x = 4$, $x = 4(a - 2)/(a + 1)$; if $a = 1$, $x \in [-2; 4]$; if $a > 1$, $x = 4$.

3.23. $a = 0$, $a = 1$.

3.24. $a \ge 2/3$.

3.25. (1) has no solutions for $k \in (-23; 0)$ and (2) has a finite nonempty set of solutions for $(-\infty; -23) \cup (0; +\infty)$.

3.26. If $a = -2$, $x \in \{-1; 15/17; 17/15\}$, if $a = -1/8$, $x \in \{-1/136; 0; 1/120\}$.

3.27. a) $a \in (-5/2; 7)$, b) $a \in [(9 - \sqrt{211})/2; -5/2] \cup \{7\}$.

3.28. $(\pi/2 + 2\pi n; 0)$, $n \in \mathbb{Z}$, $(-3\pi/2; t)$, $t \in \mathbb{R}$.

3.29. $a = 2 \pm \sqrt{2}$. **Hint:** make the substitution $b = a - 2$, $t = (x - 2)/b$.

3.30. $a \in [-8; -7] \cup [7; 8]$.

Chapter 4

Irrational Equations and Inequalities

4.1 Irrational Equations and Inequalities of the Second Degree

The equation $\sqrt{f(x)} = g(x)$ is equivalent to the system

$$\begin{cases} f(x) = g^2(x), \\ g(x) \geq 0. \end{cases}$$

Remark 4.1. The equation $\sqrt{f(x)} = g(x)$ can be solved in another way:

- square both sides of the equation $\sqrt{f(x)} = g(x)$;
- then, check the condition $g(x) \geq 0$, or substitute the found solutions into the original equation.

The inequality $\sqrt{f(x)} < g(x)$ is equivalent to the system

$$\begin{cases} f(x) < g^2(x), \\ f(x) \geq 0, \\ g(x) \geq 0. \end{cases}$$

The inequality $\sqrt{f(x)} > g(x)$ is equivalent to a set of systems

$$\left[\begin{array}{l} \begin{cases} f(x) \geq 0, \\ g(x) < 0. \end{cases} \\ \begin{cases} f(x) > g^2(x), \\ g(x) \geq 0. \end{cases} \end{array}\right.$$

Example 4.1. Solve the equation

$$\sqrt{2x + 5} - \sqrt{3x - 5} = 2. \qquad (4.1)$$

Proof. The domain of definition is determined by the inequalities $2x + 5 \geq 0$ and $3x - 5 \geq 0$, whence $x \geq 5/3$. We transform Equation (4.1) to the form

$$\sqrt{2x + 5} = 2 + \sqrt{3x - 5}.$$

Both parts of this equation are non-negative. So, on squaring the original equation, we get the equation

$$2x + 5 = 4 + 4\sqrt{3x - 5} + 3x - 5.$$

Converting it,

$$6 - x = 4\sqrt{3x - 5},$$

and squaring both parts of this equation,

$$x^2 - 12x + 36 = 16(3x - 5).$$

Therefore,

$$x^2 - 60x + 116 = 0,$$

from which we get $x_1 = 2$ and $x_2 = 58$. Both of these roots are included in the domain of definition (4.1). However, when substituting $x = 58$ into equality (4.1), we get

$$\sqrt{121} - \sqrt{169} = 2 \quad \text{or} \quad 11 - 13 = 2,$$

which is incorrect. Substituting $x = 2$ gives the correct equality $3 - 1 = 2$.

□

Example 4.2. Solve the equation

$$\sqrt{2x-6}+\sqrt{x+4}=5.$$

Proof. The domain of definition is determined by the inequality $x \geq 3$. Both parts of the equation are non-negative, so on squaring them, we get an equation equivalent to the original equation on the set $x \geq 3$:

$$2x-6+2\sqrt{2x-6}\sqrt{x+4}+x+4=25,$$

or

$$2\sqrt{2x-6}\sqrt{x+4}=27-3x. \qquad (4.2)$$

Let us square both parts of it. We get

$$4(2x-6)(x+4)=729-162x+9x^2, \qquad (4.3)$$

or

$$x^2-170x+825=0.$$

The roots of this equation are $x_1 = 5$ and $x_2 = 165$. Both of them belong to the domain of definition, but only $x_1 = 5$ satisfies the condition $27 - 3x \geq 0$ or $x \leq 9$. As noted above, the roots of Equation (4.2) are those roots of Equation (4.3), for which $27 - 3x \geq 0$.

Answer: $x = 5$. ☐

Consider the function $f(x) = \sqrt{2x-6}+\sqrt{x+4}-5$ (see Fig. 4.1) from the previous Example 4.2.

Remark 4.2. We could solve Example 4.2 in another way. Indeed, since the function $f(x)$ (see Fig. 4.2) is strictly increasing on the entire real axis, the function $f(x)$ takes each value only once. The solution $x = 5$ is easily guessed, i.e., $f(5) = 0$. Therefore, the root $f(x) = 0$ will be 5; hence, this solution is unique in view of monotony.

Example 4.3. Solve the inequality

$$\sqrt{x+1}>x-2.$$

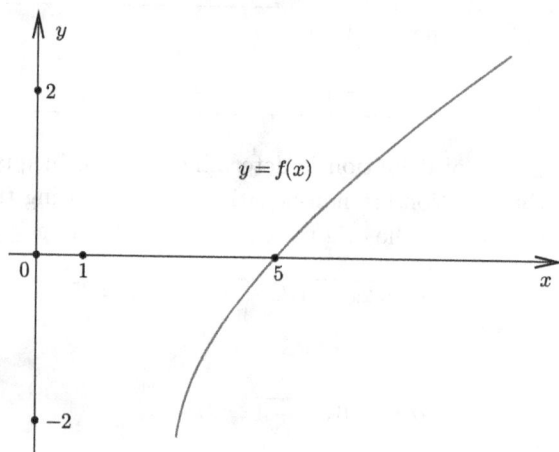

Fig. 4.1. Graph of $f(x)$.

Proof. Let us consider two cases:

1. For $x - 2 < 0$, i.e., $x < 2$, the initial inequality will be performed on the domain of definition $x \geq -1$. Intersecting these two sets, we get a partial answer: $x \in [-1; 2)$.
2. For $x - 2 \geq 0$, i.e., $x \geq 2$, the inequality is equivalent to

$$x + 1 > (x - 2)^2 \iff x^2 - 5x + 3 < 0$$
$$\iff (x - 5/2)^2 - 13/4 < 0$$
$$\iff \left(x - \frac{5 + \sqrt{13}}{2}\right) \cdot \left(x - \frac{5 - \sqrt{13}}{2}\right) < 0$$
$$\iff x \in \left(\frac{5 - \sqrt{13}}{2}; \frac{5 + \sqrt{13}}{2}\right).$$

Intersecting with the considered set, we get the second partial answer: $x \in [2; (5 + \sqrt{13})/2)$.

Now, combining the partial answers, we arrive at the final answer.

Answer: $x \in [-1; (5 + \sqrt{13})/2)$.   □

Example 4.4. Solve the inequality

$$\frac{1}{\left(\sqrt{\dfrac{2-x}{x}} - \dfrac{x+1}{2x}\right)^2} \geq 0.$$

Proof. It is easy to see that the initial inequality is fulfilled on the domain of definition. Therefore, this task is reduced to finding the domain of definition. Let us find it:

$$\begin{cases} \sqrt{\dfrac{2-x}{x}} - \dfrac{x+1}{2x} \neq 0, \\[3mm] \dfrac{2-x}{x} \geq 0, \\[3mm] x \neq 0. \end{cases}$$

Find the roots of the equation $\sqrt{\frac{2-x}{x}} - \frac{x+1}{2x} = 0$. To do this, multiplying this equation by -4, we get

$$\frac{2-x}{x} - 4\sqrt{\frac{2-x}{x}} + 3 = 0.$$

This is a quadratic equation with respect to the variable $t = \sqrt{\frac{2-x}{x}}$. Its roots are $t_{1,2} = 1, 3$. Finally,

$$\begin{cases} \sqrt{\dfrac{2-x}{x}} \neq 1, \\[3mm] \sqrt{\dfrac{2-x}{x}} \neq 3, \\[3mm] x \in (0;2]. \end{cases} \iff \begin{cases} \dfrac{2-x}{x} \neq 1, \\[3mm] \dfrac{2-x}{x} \neq 9, \\[3mm] x \in (0;2]. \end{cases} \iff \begin{cases} x \neq 1, \\[3mm] x \neq 1/5, \\[3mm] x \in (0;2]. \end{cases}$$

Answer: $x \in (0;1/5) \cup (1/5;1) \cup (1;2]$. □

Example 4.5. Solve the equation

$$\sqrt{x-2} + \sqrt{4-x} = \sqrt{6-x}.$$

Proof. The domain of definition is determined by the inequalities $x-2 \geq 0$, $4-x \geq 0$, and $6-x \geq 0$. We get $2 \leq x \leq 4$. Both parts of the equation are non-negative, so on squaring them, we get an equivalent equation

$$x - 2 + 2\sqrt{x-2}\sqrt{4-x} + 4 - x = 6 - x,$$

or

$$2\sqrt{x-2}\sqrt{4-x} = 4 - x.$$

In the domain of definition, the right side of this equation is non-negative. So, on squaring both parts, we get an equation that is equivalent to the original one in the domain of definition:

$$4(x-2)(4-x) = (4-x)^2,$$

or

$$(4-x)(5x-12) = 0.$$

This equation has the roots $x_1 = 12/5$ and $x_2 = 4$. Both of these roots are within the domain of definition.

Answer: $x_1 = 12/5$, $x_2 = 4$. □

Example 4.6. Solve the equation

$$\sqrt{2x^2 + 3x + 2} - \sqrt{7x^2 + 8} + \sqrt{2x^2 - 3x + 2} = 0.$$

Proof. First of all, we note that for all x, the expressions $2x^2 + 3x + 2$, $7x^2 + 8$, and $2x^2 - 3x + 2$ are non-negative since the discriminants of all these square trinomials are negative, and the higher coefficients are positive. We transform the equation to the form

$$\sqrt{2x^2 + 3x + 2} + \sqrt{2x^2 - 3x + 2} = \sqrt{7x^2 + 8}.$$

Both sides of the equation are non-negative. Let us square them:

$$2x^2 + 3x + 2 + 2\sqrt{2x^2 + 3x + 2}\sqrt{2x^2 - 3x + 2} + 2x^2 - 3x + 2 = 7x^2 + 8.$$

Further,

$$2\sqrt{(2x^2 + 2) + 3x}\sqrt{(2x^2 + 2) - 3x} = 3x^2 + 4,$$

or

$$2\sqrt{(2x^2 + 2)^2 - 9x^2} = 3x^2 + 4.$$

Since again both parts of the equation are non-negative, on squaring them, we get the equivalent equation

$$4((2x^2 + 2)^2 - 9x^2) = 9x^4 + 24x + 16,$$

or

$$x^4 - 4x^2 = 0.$$

This is a biquadrate equation. Its roots are $x_1 = -2$, $x_2 = 0$, and $x_3 = 2$.

Answer: $x_1 = -2$, $x_2 = 0$, $x_3 = 2$. □

Example 4.7. Solve the equation

$$\sqrt{1 + x\sqrt{x^2 - 24}} = x - 1.$$

Proof. Let us square both parts of this equation:

$$1 + x\sqrt{x^2 - 24} = x^2 - 2x + 1.$$

We transform the resulting equation to the form

$$x(\sqrt{x^2 - 24} - (x - 2)) = 0. \qquad (4.4)$$

The root $x = 0$ of the equation (4.4) does not satisfy the original equation since at $x = 0$, the value $x^2 - 24$ standing under the root is $-24 < 0$. It remains to solve the equation

$$\sqrt{x^2 - 24} = x - 2.$$

By squaring both parts of it, we find

$$x^2 - 24 = x^2 - 4x + 4,$$

which gives $x = 7$. On substituting the obtained value into the original equation, we have the equality $6 = 6$.

Answer: $x = 7$. □

4.2 Solving Irrational Equations Using Variable Substitution

The introduction of an auxiliary variable in some cases leads to a simplification of the equation. Most often, the radical included in the equation is used as a new variable. In this case, the equation becomes algebraic with respect to the new variable.

Example 4.8. Solve the equation

$$x + \sqrt{x+1} = 1.$$

Proof. Let us put $y = \sqrt{x+1}$. Then, $y^2 = x + 1$, from which we get $x = y^2 - 1$ and

$$y^2 - 1 + y = 1 \iff y^2 + y - 2 = 0.$$

The roots of this equation are $y_1 = 1$, $y_2 = -2$. It remains to solve the equations

$$\sqrt{x+1} = 1 \quad \text{and} \quad \sqrt{x+1} = -2.$$

From the first of them, we find $x = 0$, whereas the second has no solutions.
$\square$

Example 4.9. Solve the equation

$$2x^2 + 3x + \sqrt{2x^2 + 3x + 9} = 33.$$

Proof. Assuming $y = \sqrt{2x^2 + 3x + 9}$, we arrive at the equation

$$y^2 + y - 42 = 0,$$

which gives $y_1 = 6$ and $y_2 = -7$. Returning to the variable x, we obtain the equations

$$\sqrt{2x^2 + 3x + 9} = 6 \quad \text{and} \quad \sqrt{2x^2 + 3x + 9} = -7.$$

We solve the first equation by squaring both parts:

$$2x^2 + 3x + 9 = 36.$$

The roots of this equation are $x_1 = -9/2$ and $x_2 = 3$. The second equation

$$\sqrt{2x^2 + 3x + 9} = -7$$

has no solutions since the square root is always non-negative.

Answer: $x_1 = -9/2$, $x_2 = 3$. □

Example 4.10. Solve the equation

$$\sqrt{\frac{x+1}{x-1}} - \sqrt{\frac{x-1}{x+1}} = \frac{3}{2}. \qquad (4.5)$$

Proof. Denote $y = \sqrt{\frac{x+1}{x-1}}$. Then, $\sqrt{\frac{x-1}{x+1}} = \frac{1}{y}$ and Equation (4.5) turns into the equation

$$y - \frac{1}{y} = \frac{3}{2},$$

or

$$\frac{y^2 - \frac{3}{2}y - 1}{y} = 0,$$

which gives $y_1 = 2$ and $y_2 = -1/2$. It remains to solve the equations

$$\sqrt{\frac{x+1}{x-1}} = 2 \quad \text{and} \quad \sqrt{\frac{x+1}{x-1}} = -\frac{1}{2}.$$

From the first of them, after squaring both parts, we get

$$\frac{x+1}{x-1} = 4, \qquad x = \frac{5}{3}.$$

Substituting the found value $x = 5/3$ into equation (4.5), we get the correct equality. The second equation has no solutions since its left side is always non-negative.

Answer: $x = 5/3$. □

Let us look at examples in which a more complex variable replacement is used.

Example 4.11. Solve the equation

$$2x^2 + (2x+1)\sqrt{x^2 - x + 1} = 1. \tag{4.6}$$

Proof. Let us move all its members to the left part:

$$2x^2 + (2x+1)\sqrt{x^2 - x + 1} - 1 = 0,$$

and we make an additional transformation:

$$x^2 + 2x\sqrt{x^2 - x + 1} + x^2 - x + 1 + x + \sqrt{x^2 - x + 1} - 2 = 0.$$

Note that

$$x^2 + 2x\sqrt{x^2 - x + 1} + x^2 - x + 1 = (x + \sqrt{x^2 - x + 1})^2,$$

and put $y = x + \sqrt{x^2 - x + 1}$. Then, the equation takes the form

$$y^2 + y - 2 = 0,$$

so we get $y_1 = 1$, $y_2 = -2$. Let us solve the equations

$$x + \sqrt{x^2 - x + 1} = 1 \quad \text{and} \quad x + \sqrt{x^2 - x + 1} = -2,$$

or

$$\sqrt{x^2 - x + 1} = 1 - x \quad \text{and} \quad \sqrt{x^2 - x + 1} = -2 - x.$$

Squaring both parts of the first one, we get

$$x^2 - x + 1 = x^2 - 2x + 1,$$

or

$$-x = 0,$$

from which we get $x = 0$. Substituting it into the original equation, we get the correct equality $1 = 1$. Let us square both parts of the remaining equation:

$$x^2 - x + 1 = x^2 + 4x + 4.$$

Therefore, $5x = -3$, $x = -\frac{3}{5}$. By checking, we make sure that this value does not satisfy equation (4.6).

Answer: $x = 0$. □

Example 4.12. Solve the equation

$$\frac{1}{x} + \frac{1}{\sqrt{1-x^2}} = \frac{35}{12}. \tag{4.7}$$

Proof. Let us square both parts of the equation. We obtain the equation

$$\frac{1}{x^2} + \frac{2}{x\sqrt{1-x^2}} + \frac{1}{1-x^2} = \frac{1225}{144}, \tag{4.8}$$

which is a consequence of equation (4.7). Note that

$$\frac{1}{x^2} + \frac{1}{1-x^2} = \frac{1-x^2+x^2}{x^2(1-x^2)} = \frac{1}{x^2(1-x^2)} = \left(\frac{1}{x\sqrt{1-x^2}}\right)^2.$$

Denote $y = \frac{1}{x\sqrt{1-x^2}}$. We reduce equation (4.8) to the form

$$y^2 + 2y - \frac{1225}{144} = 0,$$

from which we get $y_1 = 25/12$ and $y_2 = -49/12$. It remains to solve the equations

$$\frac{1}{x\sqrt{1-x^2}} = \frac{25}{12} \quad \text{and} \quad \frac{1}{x\sqrt{1-x^2}} = -\frac{49}{12}. \tag{4.9}$$

The solutions to the first of equations (4.9) satisfy the inequality $x > 0$, and the solutions of the second inequality satisfy $x < 0$. We solve the first by squaring both of its parts:

$$\frac{1}{x^2(1-x^2)} = \frac{625}{144},$$

or

$$x^2(1-x^2) = \frac{144}{625}.$$

This is a biquadrate equation. Denoting $x^2 = y$, we get

$$y^2 - y + \frac{144}{625} = 0,$$

and $y_1 = 9/25, y_2 = 16/25$. Since $x > 0$, from these equalities, we find $x_1 = 3/5$ and $x_2 = 4/5$. The second of equations (4.9) is solved in the same way.

Answer: $x_1 = \frac{3}{5}$, $x_2 = \frac{4}{5}$, $x_3 = -\frac{5+\sqrt{73}}{14}$. $\qquad\square$

Example 4.13. Solve the equation

$$x + \frac{x}{\sqrt{x^2 - 1}} = \frac{35}{12}.$$

Proof. First, we note that if x is the solution of this equation, then $x > 0$. Since the left side of the equation can be represented as

$$x\left(1 + \frac{1}{\sqrt{x^2 - 1}}\right),$$

the quantity

$$1 + \frac{1}{\sqrt{x^2 - 1}}$$

is positive for all possible values of x. By replacing the variable $t = 1/x$ and noting that, in view of the fact that $x > 0$,

$$\frac{x}{\sqrt{x^2 - 1}} = \frac{1}{\sqrt{1 - \frac{1}{x^2}}} = \frac{1}{\sqrt{1 - t^2}}.$$

We reduce equation (4.13) to the form

$$\frac{1}{t} + \frac{1}{\sqrt{1 - t^2}} = \frac{35}{12}.$$

However, this is an equation from the previous example, which has the roots $t_1 = \frac{3}{5}$, $t_2 = \frac{4}{5}$, and $t_3 = -\frac{5+\sqrt{73}}{14}$. In view of the remark made above, the inequality $x > 0$ holds for the solutions of equation (4.13). So, $1/t_3$ is not a root of equation (4.13).

Answer: $t_1 = \frac{5}{3}, t_2 = \frac{5}{4}$. □

4.3   Solving Equations by Factoring the Expressions Included in It

Solving equations of the form

$$f(x)g(x) = 0$$

comes down to solving the equations

$$f(x) = 0, \quad g(x) = 0$$

and checking the obtained roots.

Example 4.14. Solve the equation

$$(x+3)\sqrt{x-1} = 3\sqrt{x^2-1}. \tag{4.10}$$

Proof. The domain of definition is $x \geq 1$. On this set,

$$\sqrt{x^2-1} = \sqrt{x-1}\sqrt{x+1},$$

the equation is transformed to the form

$$\sqrt{x-1}\left(x+3-3\sqrt{x+1}\right) = 0.$$

Let us solve the equations

$$\sqrt{x-1} = 0 \quad \text{and} \quad x+3-3\sqrt{x+1} = 0.$$

The root of the first equation is $x_1 = 1$. It also satisfies equation (4.10). The second equation is solved using the substitution $y = \sqrt{x+1}$. It takes the form

$$y^2 - 3y + 2 = 0.$$

The roots of this equation are $y_1 = 1$ and $y_2 = 2$. Returning to the variable x, we find $x_2 = 3$ and $x_3 = 0$. Through substitution, we make sure that $x_2 = 3$ satisfies equation (4.10), but $x_3 = 0$ does not satisfy this equation.

Answer: $x_1 = 1$, $x_2 = 3$. $\qquad\square$

Example 4.15. Solve the equation

$$\sqrt{2x^2 + 21x - 11} - \sqrt{2x^2 - 9x + 4} = \sqrt{18x - 9}.$$

Proof. The domain of definition is determined by the inequalities $2x^2 + 21x - 11 \geq 0$, $2x^2 - 9x + 4 \geq 0$, $18x - 9 \geq 0$ or

$$(2x-1)(x+11) \geq 0, \quad (2x-1)(x-4) \geq 0, \quad 9(2x-1) \geq 0,$$

which gibes $x = 1/2$ or $x \geq 4$. Using the factorizations obtained above, we transform the original equation (4.15) to the form

$$\sqrt{2x-1}\left(\sqrt{x+11} - \sqrt{x-4} - 3\right) = 0.$$

The equation

$$\sqrt{2x-1}=0$$

has a single root, $x=1/2$, satisfying equations (4.15). Consider the second equation

$$\sqrt{x+11}-\sqrt{x-4}-3=0 \Longleftrightarrow \sqrt{x+11}=\sqrt{x-4}+3.$$

Both parts of the last equation are non-negative, so on squaring them, we get the equation

$$x+11=x-4+6\sqrt{x-4}+9,$$

or

$$\sqrt{x-4}=1,$$

whose root $x=5$ satisfies equation (4.15).

Answer: $x_1=1/2$, $x_2=5$.   □

Example 4.16. Solve the equation

$$2x\sqrt{x}-4x-\sqrt{x^2+3x}+2\sqrt{x+3}=0. \qquad (4.11)$$

Proof. Let us group the first term with the second and the third with the fourth:

$$(\sqrt{x}-2)2x-\sqrt{x+3}(\sqrt{x}-2)=0.$$

Therefore, the original equation takes the form

$$(\sqrt{x}-2)(2x-\sqrt{x+3})=0.$$

Solving the equation $\sqrt{x}-2=0$, we find its root as $x_1=4$, which satisfies equation (4.11). Consider the equation

$$2x-\sqrt{x+3}=0,$$

or

$$2x=\sqrt{x+3}.$$

We solve it by squaring both its parts. The resulting equation

$$4x^2 = x + 3$$

has the roots $x_2 = 1$ and $x_3 = -3/4$. But only $x_2 = 1$ satisfies (4.11).

Answer: $x_1 = 4$, $x_2 = 1$. □

It can be difficult to allocate a common multiplier. Sometimes it is possible to do this after additional transformations. In the following example, pairwise differences of rooted expressions are considered for this purpose.

Example 4.17. Solve the equation

$$\sqrt{2x^2 - 1} + \sqrt{x^2 - 3x - 2} = \sqrt{2x^2 + 2x + 3} + \sqrt{x^2 - x + 2}. \quad (4.12)$$

Proof. By rewriting equation (4.12) as

$$\sqrt{2x^2 - 1} + \sqrt{x^2 - 3x - 2} - \sqrt{2x^2 + 2x + 3} - \sqrt{x^2 - x + 2} = 0,$$

we can note that the differences of the root expressions of the first and third as well as the second and fourth terms of this equation are equal to the same value:

$$-2x - 4.$$

Let us use the identity

$$\sqrt{a} - \sqrt{b} = \frac{a - b}{\sqrt{a} + \sqrt{b}}, \quad a, b \geq 0, \ a + b > 0,$$

and transform the equation to the form

$$\frac{-2x - 4}{\sqrt{2x^2 - 1} + \sqrt{2x^2 + 2x + 3}} + \frac{-2x - 4}{\sqrt{x^2 - x + 2} + \sqrt{x^2 - 3x - 2}} = 0,$$

or

$$(2x + 4)\left(\frac{1}{\sqrt{2x^2 - 1} + \sqrt{2x^2 + 2x + 3}} + \frac{1}{\sqrt{x^2 - x + 2} + \sqrt{x^2 - 3x - 2}}\right) = 0.$$

The root of the equation $2x + 4 = 0$, i.e., the number $x = -2$, when substituted into equations (4.12) gives the correct equality. The equation

$$\frac{1}{\sqrt{2x^2 - 1} + \sqrt{2x^2 + 2x + 3}} + \frac{1}{\sqrt{x^2 - x + 2} + \sqrt{x^2 - 3x - 2}} = 0$$

has no solutions since its left side is positive in its domain of definition.

Answer: $x = -2$. □

Example 4.18. Solve the equation

$$\sqrt{(x+2)(2x-1)} - 3\sqrt{x+6} = 4 - \sqrt{(x+6)(2x-1)} + 3\sqrt{x+2}. \quad (4.13)$$

Proof. Transform equation (4.13), keeping in mind $x \geq 1/2$:

$$\sqrt{2x-1}(\sqrt{x+2} + \sqrt{x+6}) - 3(\sqrt{x+6} + \sqrt{x+2}) = 4,$$

and let us use the equality

$$4 = (x+6) - (x+2) = (\sqrt{x+6} + \sqrt{x+2})(\sqrt{x+6} - \sqrt{x+2}).$$

Then, the equation takes the form

$$(\sqrt{x+2} + \sqrt{x+6})(\sqrt{2x-1} - 3 - \sqrt{x+6} + \sqrt{x+2}) = 0.$$

The equation

$$\sqrt{x+2} + \sqrt{x+6} = 0$$

has no roots since its left part in its domain of definition is greater than 0. Let us solve the equation

$$\sqrt{2x-1} - 3 = \sqrt{x+6} - \sqrt{x+2}.$$

Squaring both of its parts, we get

$$2x - 1 - 6\sqrt{2x-1} + 9 = x + 6 - 2\sqrt{x+6}\sqrt{x+2} + x + 2,$$

or

$$3\sqrt{2x-1} = \sqrt{x+2}\sqrt{x+6}.$$

Squaring again, we arrive at the equation

$$9(2x-1) = (x+2)(x+6),$$

or

$$x^2 - 10x + 21 = 0,$$

which has the roots $x_1 = 3$ and $x_2 = 7$. By checking, we make sure that equation (4.13) is satisfied only $x_2 = 7$.

Answer: $x_2 = 7$.   □

Example 4.19. Solve the equation

$$\sqrt{2x+4} - 2\sqrt{2-x} = \frac{12x-8}{\sqrt{9x^2+16}}. \qquad (4.14)$$

Proof. The domain of definition is determined by the inequalities

$$2x+4 \geq 0, 2-x \geq 0,$$

that is, $-2 \leq x \leq 2$. Noting that

$$12x - 8 = 2(\sqrt{2x+4} + 2\sqrt{2-x})(\sqrt{2x+4} - 2\sqrt{2-x}),$$

transform equation (4.14) as follows:

$$\sqrt{2x+4} - 2\sqrt{2-x} = (\sqrt{2x+4} - 2\sqrt{2-x})\frac{2(\sqrt{2x+4} + 2\sqrt{2-x})}{\sqrt{9x^2+16}},$$

or

$$(\sqrt{2x+4} - 2\sqrt{2-x})\left(1 - \frac{(\sqrt{2x+4} + 2\sqrt{2-x})}{\sqrt{9x^2+16}}\right) = 0.$$

The equation

$$\sqrt{2x+4} = 2\sqrt{2-x}$$

is solved by squaring both its parts:

$$2x + 4 = 4(2-x),$$

and we get $x = 2/3$. The second equation we first transform to an equivalent form:

$$\sqrt{9x^2+16} = 2(\sqrt{2x+4} + 2\sqrt{2-x}).$$

Then, we square both its parts of it:

$$9x^2 + 16 = 4(12 - 2x + 4\sqrt{8 - 2x^2}),$$

and consistently transform:

$$4(8 - 2x^2) + 16\sqrt{8 - 2x^2} - (x^2 + 8x) = 0$$

$$\Longleftrightarrow 4(8 - 2x^2) + 16\sqrt{8 - 2x^2} + 16 - (x^2 + 8x + 16) = 0$$

$$\Longleftrightarrow (2\sqrt{8 - 2x^2} + 4)^2 - (x + 4)^2 = 0$$

$$\Longleftrightarrow (2\sqrt{8 - 2x^2} - x)(2\sqrt{8 - 2x^2} + x + 8) = 0.$$

The equation

$$2\sqrt{8 - 2x^2} = x \tag{4.15}$$

after squaring its parts gives

$$4(8 - 2x^2) = x^2,$$

or $9x^2 = 32$, from which we get $x_{1,2} = 4\sqrt{2}/3$. But $x_2 = -4\sqrt{2}/3$ is an extraneous root of equation (4.15). It remains to solve the equation

$$2\sqrt{8 - 2x^2} + x + 8 = 0. \tag{4.16}$$

According to the domain of definition of the original equation, $x \geq -2$. Therefore, $x + 8 \geq 6$ and

$$2\sqrt{8 - 2x^2} + x + 8 \geq 6.$$

Thus, equation (4.16) gives no solutions to the original equation.

Answer: $x = 4\sqrt{2}/3$. $\qquad\qquad\qquad\qquad\qquad\qquad\qquad$ □

Example 4.20. Solve the equation

$$\sqrt{12 - \frac{12}{x^2} - x^2} + \sqrt{x^2 - \frac{12}{x^2}} = 0. \tag{4.17}$$

Proof. Let us transform it first as follows:

$$\sqrt{12 - \frac{12}{x^2}} = x^2 - \sqrt{x^2 - \frac{12}{x^2}},$$

and we square both parts of the resulting equation:

$$12 - \frac{12}{x^2} = x^4 - 2x^2\sqrt{x^2 - \frac{12}{x^2}} + x^2 - \frac{12}{x^2},$$

from which we get

$$x^4 - 12 - 2|x|\sqrt{x^4 - 12} + x^2 = 0,$$

or

$$(\sqrt{x^4 - 12} - |x|)^2 = 0.$$

Therefore,

$$\sqrt{x^4 - 12} = |x|,$$

or

$$x^4 - 12 = x^2.$$

Solving this biquadrate equation, we obtain the equations

$$x^2 = 4 \quad \text{and} \quad x^2 = -3.$$

The first of them has the roots $x_1 = -2$ and $x_2 = 2$, satisfying equation (4.17), and the second has no solutions.

Answer: $x_1 = -2$, $x_2 = 2$.   □

4.4 Irrational Equations and Inequalities Containing Roots of Degree Higher than the Second

If the equation has the form

$$\sqrt[n]{f(x)} = g(x), \tag{4.18}$$

then it can be solved by raising both parts of this equation to the power of n. The resulting equation

$$f(x) = (g(x))^n$$

for odd n is equivalent to equation (4.18), and for even n, it is its consequence.

Example 4.21. Solve the equation

$$\sqrt[3]{x^3 + 1} = x + 1.$$

Proof. Having raised both its parts to the cubic power, we obtain the equation

$$x^3 + 1 = x^3 + 3x^2 + 3x + 1,$$

or

$$3x^2 + 3x = 0,$$

which has the roots $x_1 = -1$ and $x_2 = 0$.

Answer: $x_1 = -1$, $x_2 = 0$. □

Example 4.22. Solve the equation

$$\sqrt[3]{2 - x} = 1 - \sqrt{x - 1}. \tag{4.19}$$

Proof. By cubing both sides of equation (4.19), we get

$$2 - x = 1 - 3\sqrt{x - 1} + 3(x - 1) - (x - 1)\sqrt{x - 1},$$

or

$$\sqrt{x - 1}(x + 2) - 4x + 4 = 0,$$

which gives

$$\sqrt{x - 1}(x + 2 - 4\sqrt{x - 1}) = 0.$$

Solving the equation

$$\sqrt{x - 1} = 0,$$

we get $x_1 = 1$. This value satisfies equation (4.19). We solve the equation

$$x + 2 = 4\sqrt{x - 1}$$

by squaring both its parts:

$$x^2 + 4x + 4 = 16(x - 1).$$

The resulting equation

$$x^2 - 12x + 20 = 0$$

has the roots $x_2 = 2$ and $x_3 = 10$, which satisfy equality (4.19).

Answer: $x_1 = 1$, $x_2 = 2$, $x_3 = 10$. □

Example 4.23. Solve the equation

$$\sqrt[3]{x+1} + \sqrt[3]{x+2} + \sqrt[3]{x+3} = 0.$$

Proof. We transform the equations into the form

$$\sqrt[3]{x+1} + \sqrt[3]{x+3} = -\sqrt[3]{x+2}. \tag{4.20}$$

Let us cube both parts of the resulting equation:

$$x+1+3\sqrt[3]{(x+1)^2}\sqrt{x+3} + 3\sqrt{x+1}\sqrt[3]{(x+3)^2} + x + 3 = -x - 2.$$

Grouping, we get

$$3(x+2) + 3\sqrt[3]{(x+1)(x+3)}(\sqrt[3]{x+1} + \sqrt[3]{x+3}) = 0.$$

Using equality (4.20), we have

$$x+2+\sqrt[3]{(x+1)(x+3)}(-\sqrt[3]{x+2}) = 0,$$

or

$$\sqrt[3]{x+2}(\sqrt[3]{(x+1)(x+3)} - \sqrt[3]{(x+2)^2}) = 0.$$

The equation

$$\sqrt[3]{x+2} = 0$$

has the root $x = -2$. By cubing both sides of the equation

$$\sqrt[3]{(x+1)(x+3)} = \sqrt[3]{(x+2)^2},$$

we get the equation

$$x^2 + 4x + 3 = x^2 + 4x + 4,$$

which has no solutions.

Answer: $x = -2$. $\qquad\square$

Let us show that Example 4.23 can be solved similarly to Example 4.2. Consider the function $f(x) = \sqrt[3]{x+1} + \sqrt[3]{x+2} + \sqrt[3]{x+3}$ (see Fig. 4.2) from the previous Example 4.23.

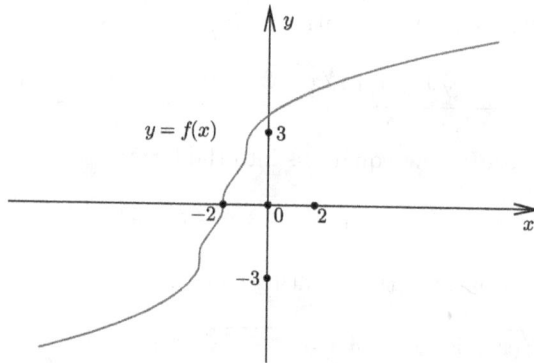

Fig. 4.2.   Graph of $f(x)$.

Remark 4.3. We could solve Example 4.23 in another way. Indeed, since the function $f(x)$ (see Fig. 4.2) is strictly increasing on the entire real axis, the function $f(x)$ takes each value only once. The solution $x = -2$ is easily guessed, i.e., $f(-2) = 0$. Therefore, the root of the equation $f(x) = 0$ will be -2, hence this solution is unique in view of monotony.

Example 4.24. Solve the equation

$$x + \sqrt[8]{x^5} - 12\sqrt[4]{x} = 0. \tag{4.21}$$

Proof. The domain of definition is determined by the inequality $x \geq 0$. Let us transform equation (4.21) by taking the multiplier $\sqrt[4]{x}$ out of the brackets on the left side:

$$\sqrt[4]{x}(\sqrt[4]{x^3} + \sqrt[8]{x^3} - 12) = 0.$$

One root of this equation is $x_1 = 0$. To solve the equation

$$\sqrt[4]{x^3} + \sqrt[8]{x^3} - 12 = 0, \tag{4.22}$$

let us put $y = \sqrt[8]{x^3}$. Then, equation (4.22) takes the form

$$y^2 + y - 12 = 0.$$

The roots of this equation are $y_1 = 3$ and $y_2 = -4$. It remains to solve the equations

$$\sqrt[8]{x^3} = 3 \quad \text{and} \quad \sqrt[8]{x^3} = -4.$$

The second of these equations has no solutions since $\sqrt[8]{x^3} \geq 0$, for all x, in the domain of definition. Solving the first equation, we get $x_2 = 3^{8/3}$.

Answer: $x_1 = 0$, $x_2 = 3^{8/3}$. ☐

Example 4.25. Solve the equation

$$\sqrt[3]{12 - x} + \sqrt[3]{14 + x} = 2. \tag{4.23}$$

Proof. Let us put $12 - x = u^3$ and $14 + x = v^3$. Then, $u^3 + v^3 = 26$, and equation (4.23) takes the form $u + v = 2$. Thus,

$$\begin{cases} u + v = 2, \\ u^3 + v^3 = 26. \end{cases}$$

Using the identity

$$u^3 + v^3 = (u + v)(u^2 + uv + v^2),$$

bearing in mind the first equation of the system, we obtain

$$2(u^2 - uv + v^2) = 26,$$

or

$$u^2 - u(2 - u) + (2 - u)^2 = 13,$$

or

$$u^2 - 2u - 3 = 0,$$

which gives $u_1 = 3$ and $u_2 = -1$. Returning to the variable x, we find $x_1 = -15$ and $x_2 = 13$.

Answer: $x_1 = -15$, $x_2 = 13$. ☐

Example 4.26. Solve the equation

$$\sqrt[3]{(8 - x)^2} + 2\sqrt[3]{(x + 27)^2} = 3\sqrt[3]{(8 - x)(x + 27)}. \tag{4.24}$$

Proof. Let us put $u = \sqrt[3]{8 - x}$, $v = \sqrt[3]{x + 27}$. Then, equation (4.24) will take the form

$$u^2 + 2v^2 = 3uv. \tag{4.25}$$

Since $x = -27$, at which the variable v becomes 0, is not a solution to equation (4.24) (as can be seen by substitution), we divide both parts of equation (4.25) by v^2:

$$\left(\frac{u}{v}\right)^2 - 3\left(\frac{u}{v}\right) + 2 = 0.$$

From this equation, we find

$$\frac{u}{v} = 1 \quad \text{and} \quad \frac{u}{v} = 2.$$

It remains to solve the equations

$$\sqrt[3]{\frac{8-x}{x+27}} = 1 \quad \text{and} \quad \sqrt[3]{\frac{8-x}{x+27}} = 2.$$

We cube both parts of each of them to get

$$\frac{8-x}{x+27} = 1 \quad \text{and} \quad \frac{8-x}{x+27} = 8,$$

which gives $x_1 = -19/2$ and $x_2 = 208/9$.

Answer: $x_1 = -19/2$, $x_2 = 208/9$.   □

Example 4.27. Solve the equation

$$\sqrt{4x - x^2} + \sqrt{4x - x^2 - 3} = 3 + \sqrt{2x - x^2}.$$

Proof. The initial equation is presented in the following form:

$$\sqrt{4 - (x-2)^2} + \sqrt{1 - (x-2)^2} = 3 + \sqrt{x(2-x)}.$$

The domain of definition is determined by the inequalities

$$\begin{cases} 4 - (x-2)^2 \geq 0, \\ 1 - (x-2)^2 \geq 0, \\ x(2-x) \geq 0. \end{cases} \iff \begin{cases} x \in [0; 4], \\ x \in [1; 3], \\ x \in [0; 2]. \end{cases} \iff x \in [1; 2].$$

We denote $f(x) = \sqrt{4 - (x-2)^2} + \sqrt{1 - (x-2)^2}$, $g(x) = 3 + \sqrt{x(2-x)}$. Our task is to solve the equation

$$f(x) = g(x), \quad x \in [1; 2].$$

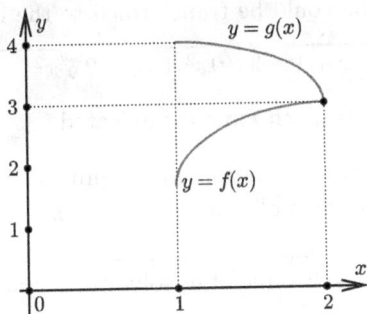

Fig. 4.3.   Graph of $f(x)$ and $g(x)$.

Note that the functions $f(x)$ and $g(x)$ are monotonic on the domain of definition (see Fig. 4.3), where the function $f(x)$ is increasing and the function $g(x)$ is decreasing. Therefore,

$$f(x) \le \sqrt{4} + \sqrt{1} = 3 \le g(x), \quad x \in [1; 2].$$

Moreover, the equal sign is achieved, due to monotony, only at one point: $x = 2$. $\square$

Example 4.28. Solve the equation

$$(2x + 1)(2 + \sqrt{(2x + 1)^2 + 3}) + 3x(2 + \sqrt{9x^2 + 3}) = 0.$$

Proof. Consider the function

$$f(t) = t(2 + \sqrt{t^2 + 3}).$$

The equation takes the form

$$f(2x + 1) + f(3x) = 0 \Longleftrightarrow f(2x + 1) = -f(3x).$$

Since the function $f(t)$ is odd, this equation takes the form

$$f(2x + 1) = f(-3x).$$

Since $f(t)$ is an increasing function, the equation is equivalent to the equation

$$2x + 1 = -3x \Longleftrightarrow x = -\frac{1}{5}.$$

Note that this equation could be transformed to the form

$$x(10 + 4\sqrt{x^2 + x + 1} + 3\sqrt{9x^2 + 3}) + 2\sqrt{x^2 + x + 1} + 2 = 0.$$

This last equation looks much more complicated. □

Similarly, the following equation will require making several transformations that will greatly simplify its appearance.

Example 4.29. For each value of a, solve the equation

$$\sqrt{-x^3 + (a-1)x^2 + (a-1)x + a} = 2x^2 + 3x + 2 - a.$$

Proof. The presence of a parameter in the equation, in this case, simplifies rather than complicates the solution. Namely, it is immediately clear that $x = a$ is the root of the polynomial standing under the root. Therefore,

$$\sqrt{-x^3 + (a-1)x^2 + (a-1)x + a}$$
$$= \sqrt{(x-a)(-x^2 - x - 1)} = \sqrt{(x^2 + x + 1)(a - x)}.$$

Since $x^2 + x + 1 > 0$, for all x, the equation is defined under the condition $a - x \geq 0$. It remains to be noted that $2x^2 + 3x + 2 - a = 2(x^2 + x + 1) - (a - x)$. Denote $u = \sqrt{x^2 + x + 1}$, $v = \sqrt{a - x}$ and get the equation

$$uv = 2u^2 - v^2, \quad 2u^2 - uv - v^2 = 0,$$

or

$$(u - v)(2u + v) = 0.$$

We return to the variable x and get

$$\sqrt{x^2 + x + 1} = \sqrt{a - x},$$

and

$$x^2 + x + 1 = a - x \iff (x+1)^2 = a,$$

from which we get $x = -1 \pm \sqrt{a}$, for $a \geq 0$. Both of these roots satisfy the condition $a \geq x$. If $a < 0$, then there are no solutions. The other equation,

$$2\sqrt{x^2 + x + 1} = -\sqrt{a - x},$$

can have solutions only if $x^2 + x + 1 = 0$ and $a - x = 0$ simultaneously. But the first of these equations has no roots. □

The considered example can serve as a model for constructing very difficult tasks.

Example 4.30. Solve the inequality

$$\sqrt{2x^2 - 1} + |x| \leq x^2 - 1.$$

(This example will appear again when solving logarithmic inequalities.)

Proof. Here, a less obvious transformation can also help us solve the problem:

$$x^2 - 1 = 2x^2 - 1 - x^2 = (\sqrt{2x^2 - 1} - |x|)(\sqrt{2x^2 - 1} + |x|),$$

where the original inequality in the domain of definition $|x| \geq \frac{\sqrt{2}}{2}$ can be transformed to the form

$$(\sqrt{2x^2 - 1} + |x|) \leq (\sqrt{2x^2 - 1} - |x|)(\sqrt{2x^2 - 1} + |x|),$$

or to the equivalent form

$$1 \leq \sqrt{2x^2 - 1} - |x| \Longleftrightarrow |x| + 1 \leq \sqrt{2x^2 - 1},$$

and

$$x^2 + 2|x| + 1 \leq 2x^2 - 1 \Longleftrightarrow x^2 - 2|x| - 2 \geq 0,$$

from which we get

$$|x| \geq 1 + \sqrt{3}. \qquad \square$$

4.5 Trigonometric Substitutions in Irrational Equations

Example 4.31. Solve the equation

$$\sqrt{1 - x^2}(1 - 4x^2) + x(3 - 4x^2) = \sqrt{2}.$$

Proof. The domain of definition for the equation is given by the inequalities $-1 \leq x \leq 1$. The expression $3x - 4x^3$ when substituting $x = \sin t$ takes the form $3\sin t - 4\sin^3 t = \sin 3t$. At the same time, if we assume that $x = \sin t$, $-\frac{\pi}{2} \le t \le \frac{\pi}{2}$, then the value of $\sqrt{1-x^2} = \sqrt{1-\sin^2 t} = \sqrt{\cos^2 t} = |\cos t| = \cos t$. Then,

$$1 - 4\sin^2 t = 1 - 4(1 - \cos^2 t) = -3 + 4\cos^2 t,$$

and

$$\cos t(-3 + 4\cos^2 t) = 4\cos^3 t - 3\cos t = \cos 3t.$$

So, the equation takes the form

$$\cos 3t + \sin 3t = \sqrt{2}$$

$$\Longleftrightarrow \sqrt{2}\sin\left(3t + \frac{\pi}{4}\right) = \sqrt{2} \Longleftrightarrow \sin\left(3t + \frac{\pi}{4}\right) = 1$$

$$\Longleftrightarrow 3t + \frac{\pi}{4} = \frac{\pi}{2} + 2\pi k \Longleftrightarrow t = \frac{\pi}{12} + \frac{2\pi k}{3}, \quad k \in \mathbb{Z}.$$

The condition $-\frac{\pi}{2} \le t \le \frac{\pi}{2}$ is satisfied only by $t = \frac{\pi}{12}$. In this case, $x = \sin\frac{\pi}{12} = \frac{\sqrt{6}-\sqrt{2}}{4}$. □

Remark 4.4. Note that the equation considered in Example 4.13,

$$\frac{1}{x} + \frac{1}{\sqrt{1-x^2}} = \frac{35}{12},$$

through the substitution $x = \sin t$, $-\frac{\pi}{2} < t < 0$ or $0 < t < \frac{\pi}{2}$ can be reduced to the equations

$$\frac{1}{\sin t} + \frac{1}{\cos t} = \frac{35}{12},$$

$$\sin t + \cos t = \frac{35}{12}\sin t \cos t.$$

In the following, will solve this trigonometric equation.

Using the equality $\sin t \cos t = \frac{(\sin t + \cos t)^2 - 1}{2}$, this equation reduces to a square with respect to the variable $z = \sin t + \cos t$. Such equations are discussed in detail in the chapter on trigonometric equations. Here, we get

$$35z^2 - 24z - 35 = 0,$$

$$z_1 = -\frac{5}{7} \quad \text{and} \quad z_2 = \frac{7}{5}.$$

In other words,

$$\sin t + \cos t = -\frac{5}{7} \quad \text{and} \quad \sin t + \cos t = \frac{7}{5}.$$

To solve equations of this type, we use the formula $\sin t + \cos t = \sqrt{2}\sin\left(t + \frac{\pi}{4}\right)$. Then, the first equation gives

$$t + \frac{\pi}{4} = -\arcsin\frac{5}{7\sqrt{2}} + 2\pi k, \quad k \in \mathbb{Z},$$

or

$$t + \frac{\pi}{4} = \pi + \arcsin\frac{5}{7\sqrt{2}} + 2\pi l, \quad l \in \mathbb{Z}.$$

The values of t obtained from the first of these equalities are included in the interval $\left(-\frac{\pi}{2}, 0\right)$, for $k = 0$. The value of t obtained from the second interval is not included in any of the intervals $\left(-\frac{\pi}{2}, 0\right)$, $\left(0, \frac{\pi}{2}\right)$. It turns out that $x = \sin t = -\frac{5+\sqrt{73}}{2}$.

Similarly, the second equation

$$\sqrt{2}\sin\left(t + \frac{\pi}{4}\right) = \frac{7}{5}$$

gives the equations

$$t + \frac{\pi}{4} = \arcsin\frac{7}{5\sqrt{2}} + 2\pi k, \quad k \in \mathbb{Z},$$

or

$$t + \frac{\pi}{4} = \pi - \arcsin\frac{7}{5\sqrt{2}} + 2\pi l, \quad l \in \mathbb{Z}.$$

These values of t correspond to, as before, $x = \frac{3}{5}$ and $x = \frac{4}{5}$.

Problems

4.1. Solve the inequality $\sqrt{x+3} > 5 - 2x$.

4.2. Solve the inequality $\sqrt{x+4} > x + 1$.

4.3. Solve the equation $\sqrt{35 - 5x} = 9 - 2x$.

4.4. Solve the equation $\sqrt{x+2} + \sqrt{8-x} = \sqrt{15}$.

4.5. Solve the inequality $\sqrt{5 - 4x - x^2} \geq -2x - 1$.

4.6. Solve the inequality $\sqrt{1-x} - \sqrt{x} > \dfrac{1}{\sqrt{3}}$.

4.7. Solve the equation

$$\sqrt{x^2 + 5x + 4} - \sqrt{x^2 - x - 6} = -\sqrt{2x^2 + 4x - 2}.$$

4.8. Solve the inequality $2x - 11 < 2\sqrt{36 - x^2}$.

4.9. Solve the inequality $6 \cdot \sqrt{\dfrac{2x-1}{x}} \geq \dfrac{10x-1}{x}$.

4.10. Solve the equation $\sqrt{2x^2 - 21x + 4} = 2 - 11x$.

4.11. Solve the equation $\sqrt{8x^2 - 40x + 50} = 7 - 3x$.

4.12. Solve the inequality $\dfrac{\sqrt{51 - 2x - x^2}}{1 - x} < 1$.

4.13. Solve the inequality $\dfrac{3x - 2}{\sqrt{5x - 2}} < 1$.

4.14. Solve the inequality $\dfrac{\sqrt{x^2 - 6x - 7}}{x - 7} \geq \dfrac{x+1}{3}$.

4.15. Solve the inequality $\dfrac{\sqrt{4x + 7} - 3x + 5}{16 - 3x^2 + 22x} < 0$.

4.16. Solve the equation

$$\sqrt{6x - x^2 - 5} + \sqrt{6x - x^2 - 8} = 3 + \sqrt{4x - x^2 - 3}.$$

4.17. Solve the equation $\dfrac{1}{x} + \dfrac{1}{\sqrt{1 - x^2}} = \dfrac{221}{60}$.

4.18. Solve the equation

$$\sqrt{1-x^2}(1-4x^2) + x(3-4x^2) + \sqrt{2} = 0.$$

4.19. Solve the equation

$$\sqrt{1-x^2}(1-4x^2) - \sqrt{3/2} = x(4x^2 - 3).$$

Hints and Answers

4.1. $((21 - \sqrt{89})/8; +\infty)$.

4.2. $x \in [-4; (\sqrt{13} - 1)/2)$.

4.3. 2.

4.4. $3 - 5\sqrt{3}/2$.

4.5. $[-2; 1]$.

4.6. $[0; (3 - \sqrt{5})/6)$.

4.7. -4.

4.8. $[-6; (11 + \sqrt{167})/4)$.

4.9. $x \in [-1/2; -1/14)$.

4.10. 0.

4.11. $x = 1 - \sqrt{2}$.

4.12. $[-1 - \sqrt{52}; -5) \cup (1; -1 + \sqrt{52}]$.

4.13. $(2/5; (17 + \sqrt{73})/18)$.

4.14. $(-\infty; -2] \cup \{-1\} \cup (7; 8]$.

4.15. $[-7/4; -2/3) \cup [(17 + \sqrt{127})/9; 8)$.

4.16. 3.

4.17. 5/13, 12/13. **Hint:** Make a trigonometric substitution.

4.18. $-1/\sqrt{2}$.

4.19. $\sin(\pi/36)$, $\sin(5\pi/36)$.

Chapter 5

Trigonometric Equations and Inequalities

5.1 The Simplest Trigonometric Equations

Here, we assume that the reader already knows the simplest trigonometric formulas, such as the basic trigonometric identity $\sin^2 x + \cos^2 x = 1$.

Example 5.1. Solve the equation

$$5 + \cos 2x = 6 \cos x.$$

Proof. Using the formula of the cosine of the double angle, we get

$$5 + \cos 2x = 6 \cos x \Longleftrightarrow 4 + 2 \cos^2 x = 6 \cos x$$

$$\Longleftrightarrow \cos^2 x - 3 \cos x + 2 = 0 \Longleftrightarrow (\cos x - 1)(\cos x - 2) = 0.$$

Since the equation $\cos x = 2$ has no solutions, the original equation is equivalent to $\cos x = 1$.

Answer: $2\pi n$, $n \in \mathbb{Z}$. $\qquad\qquad\qquad\qquad\qquad\qquad \square$

Example 5.2. Solve the equation

$$8 \cos 2x + 16 \cos x + 7 = 0.$$

Proof. Transform the original equation

$$8\cos 2x + 16\cos x + 7 = 0 \Longleftrightarrow 16\cos^2 x + 16\cos x - 1 = 0$$

$$\Longleftrightarrow (4\cos x + 2)^2 - 5 = 0 \Longleftrightarrow \cos x = (-2 \pm \sqrt{5})/4.$$

The equation $\cos x = -(2+\sqrt{5})/4$ has no solutions because $-(2+\sqrt{5})/4 < -1$. Solving the second equation $\cos x = (-2+\sqrt{5})/4$, we arrive at the answer $\pm\arccos((\sqrt{5}-2)/4) + 2\pi n$, $n \in \mathbb{Z}$.

Answer: $\pm\arccos((\sqrt{5}-2)/4) + 2\pi n$, $n \in \mathbb{Z}$. $\qquad\square$

Example 5.3. Solve the equation

$$2\sin^2 x + \sin^2 2x = \frac{5}{4} - 2\cos 2x.$$

Proof. We use the identities

$$2\sin^2 x = 1 - \cos 2x \quad \text{and} \quad \sin^2 2x = 1 - \cos^2 2x$$

to transform the original equation:

$$1 - \cos 2x + 1 - \cos^2 2x = \frac{5}{4} - 2\cos 2x,$$

or

$$\cos^2 2x + \cos 2x - \frac{3}{4} = 0.$$

Replace $\cos 2x$ with t:

$$t^2 + t - \frac{3}{4} = 0.$$

The roots of this equation are $t_1 = -1/2$, $t_2 = 3/2$, from which we have

$$\cos 2x = -\frac{1}{2} \quad \text{and} \quad \cos 2x = \frac{3}{2}.$$

The second equation has no solutions, and the first one gives

$$2x = \pm\frac{2\pi}{3} + 2\pi l, \quad l \in \mathbb{Z},$$

$$x = \pm\frac{\pi}{3} + \pi l, \quad l \in \mathbb{Z}.$$

Answer: $x = \pm\pi/3 + \pi l$, $l \in \mathbb{Z}$. $\qquad\square$

The following formulas will be useful:

5.1.1 Formulas for converting the sum and difference of trigonometric functions into products

$$\sin x \pm \sin y = 2\sin \frac{x \pm y}{2} \cos \frac{x \mp y}{2};$$

$$\cos x + \cos y = 2\cos \frac{x + y}{2} \cos \frac{x - y}{2};$$

$$\cos x - \cos y = -2\sin \frac{x + y}{2} \sin \frac{x - y}{2};$$

$$\tan x \pm \tan y = \frac{\sin(x \pm y)}{\cos x \cos y}, \quad x, y \neq \pi(2n+1)/2, \quad n \in \mathbb{Z};$$

$$\cot x \pm \cot y = \frac{\sin(y \pm x)}{\sin x \sin y}, \quad x, y \neq \pi n, \quad n \in \mathbb{Z};$$

$$\sin x + \cos y = \sin x + \sin\left(\frac{\pi}{2} - y\right) = 2\sin \frac{x - y + \pi/2}{2} \cos \frac{x + y - \pi/2}{2};$$

$$\sin x - \cos y = \sin x - \sin\left(\frac{\pi}{2} - y\right) = 2\sin \frac{x + y - \pi/2}{2} \cos \frac{x - y + \pi/2}{2}.$$

5.1.2 Formulas for converting the product of trigonometric functions into a sum

$$\sin x \sin y = \frac{1}{2}(\cos(x - y) - \cos(x + y));$$

$$\cos x \cos y = \frac{1}{2}(\cos(x + y) + \cos(x - y));$$

$$\sin x \cos y = \frac{1}{2}(\sin(x + y) + \sin(x - y)).$$

Remark 5.1. When solving trigonometric equations, problems are often reduced to the simplest equations of the form $\sin x = \sin y$, $\cos x = \cos y$, or $\tan x = \tan y$. These equations are equivalent to the following:

$$\sin x = \sin y \iff \begin{bmatrix} x = y + 2\pi k, & k \in \mathbb{Z} \\ x = \pi - y + 2\pi l, & l \in \mathbb{Z}. \end{bmatrix}$$

$$\cos x = \cos y \iff x = \pm y + 2\pi k, \quad k \in \mathbb{Z}.$$

$$\tan x = \tan y \Longleftrightarrow x = y + \pi k, \quad x \neq \pi/2 + \pi n,$$

$$y \neq \pi/2 + \pi m, \quad k, n, m \in \mathbb{Z}.$$

$$\cot x = \cot y \Longleftrightarrow x = y + \pi k, \quad x \neq \pi n, \quad y \neq \pi m, \quad k, n, m \in \mathbb{Z}.$$

The formulas written out above are easily obtained from formulas of the form $\sin x - \sin y = 2 \sin \dfrac{x - y}{2} \cos \dfrac{x + y}{2}$.

Example 5.4. Solve the equation

$$2 \sin(3x/2) \cos(x/2) = \sin x.$$

Proof. We use the formula of the doubled argument for the sine on the right side of the equation

$$2 \cos(x/2) \sin(3x/2) = 2 \cos(x/2) \sin(x/2)$$

$$\Longleftrightarrow \cos(x/2) \left(\sin(3x/2) - \sin(x/2) \right) = 0.$$

From the above, $\cos(x/2) = 0$ or $2 \cos x \sin(x/2) = 0$ (here, we could use the formula from Remark 5.1). Solving these equations, we find

Answer: $\pi n/2, \, n \in \mathbb{Z}$. $\qquad\qquad\qquad\qquad\qquad\qquad\qquad\qquad\square$

Example 5.5. Solve the equation

$$\cos 3x + \sin x \sin 2x = 0.$$

Proof. By representing $3x = x + 2x$ under the cosine sign, we transform the equation:

$$(\cos x \cos 2x - \sin x \sin 2x) + \sin x \sin 2x = 0$$

$$\Longleftrightarrow \cos x \cos 2x = 0.$$

Solving this equation, we find $x = \pi/2 + \pi m, \, x = \pi/4 + \pi n/2, \, m, n \in \mathbb{Z}$. $\quad\square$

Example 5.6. Solve the equation

$$\cos 2x = 2(\cos x + \sin x).$$

Proof. Transform the original equation:

$$\cos^2 x - \sin^2 x = 2(\cos x + \sin x)$$

$$\Longleftrightarrow (\cos x - \sin x - 2)(\cos x + \sin x) = 0.$$

The equation $\cos x - \sin x - 2 = 0$ has no solutions. In face, $\cos x$ and $\sin x$ are modulo limited to one, and the equality $\cos x - \sin x = 2$ means $\cos x = 1$ and $\sin x = -1$, which is impossible due to the basic trigonometric identity $\sin^2 x + \cos^2 x = 1$.

Now, we solve the second equation:

$$\cos x + \sin x = 0 \Longleftrightarrow \sin x = -\cos x$$

$$\Longleftrightarrow \tan x = -1 \Longleftrightarrow x = -\pi/4 + \pi n, \quad n \in \mathbb{Z}.$$

Answer: $-\pi/4 + \pi n$, $n \in \mathbb{Z}$. $\qquad\square$

Example 5.7. Solve the equation

$$\cos 0.2x - \cos 0.8x + \cos 0.6x = 1.$$

Proof. Let us group the terms:

$$(1 + \cos 0.8x) - (\cos 0.2x + \cos 0.6x) = 0.$$

Using the cosine sum formula, $\cos \alpha + \cos \beta = 2 \cos \left(\frac{\alpha+\beta}{2} \right) \cdot \cos \left(\frac{\alpha-\beta}{2} \right)$, we get

$$2 \cos^2 0.4x - 2 \cos 0.4x \cos 0.2x = 0.$$

Further,

$$2 \cos 0.4x (\cos 0.4x - \cos 0.2x) = 0$$

$$\Longleftrightarrow \cos 0.4x \sin 0.3x \sin 0.1x = 0.$$

Since the roots of the equation $\sin \alpha = 0$ are the roots of the equation $\sin 3\alpha = 0$, the original equation is equivalent to

$$\cos 0.4x \sin 0.3x = 0.$$

Let us solve it:

$$\left[\begin{array}{l} 0.4x = \pi/2 + \pi m, \quad m \in \mathbb{Z}, \\ 0.3x = \pi n, \quad n \in \mathbb{Z}. \end{array} \right. \Longleftrightarrow \left[\begin{array}{l} x = 5\pi/4 + 5\pi m/2, \quad m \in \mathbb{Z}, \\ x = 10\pi n/3, \quad n \in \mathbb{Z}. \end{array} \right.$$

Answer: $5\pi/4 + 5\pi m/2$, $10\pi n/3$, $m, n \in \mathbb{Z}$. $\qquad\square$

Example 5.8. Solve the equation

$$\tan(x + \pi/4) + \tan(x - \pi/4) = \tan x.$$

Proof. Note that the equality is valid:

$$\tan(x + \alpha) + \tan(x + \beta) = \frac{\sin(2x + \alpha + \beta)}{\cos(x + \alpha)\cos(x + \beta)}$$

$$= \frac{2\sin(2x + \alpha + \beta)}{\cos(2x + \alpha + \beta) + \cos(\alpha - \beta)}.$$

Now, we transform the left expression into the equation

$$\frac{2\sin 2x}{\cos 2x} = \frac{\sin x}{\cos x} \Longleftrightarrow \frac{4\sin x \cos x}{2\cos^2 x - 1} - \frac{\sin x}{\cos x} = 0$$

$$\Longleftrightarrow \frac{\sin x(4\cos^2 x - 2\cos^2 x + 1)}{(2\cos^2 x - 1)\cos x} = 0$$

$$\Longleftrightarrow \frac{\sin x(2\cos^2 x + 1)}{(2\cos^2 x - 1)\cos x} = 0 \Longleftrightarrow \sin x = 0.$$

The last transition is equivalent since, s if $\sin x = 0$, then $2\cos^2 x - 1 = 1 \neq 0$ and $\cos x \neq 0$.

Answer: πn, $n \in \mathbb{Z}$. □

5.2   Trigonometric Equations and Inequalities: Method of the Auxiliary Argument

Recall the *method of the auxiliary argument*, which consists of introducing an additional angle to simplify the expression. We demonstrate the auxiliary angle method using the example of a trigonometric equation:

$$a\cos x + b\sin x = c, \quad a^2 + b^2 \neq 0,$$

$$\Longleftrightarrow \frac{a}{\sqrt{a^2 + b^2}}\cos x + \frac{b}{\sqrt{a^2 + b^2}}\sin x = \frac{c}{\sqrt{a^2 + b^2}}.$$

The numbers $A = a/(\sqrt{a^2 + b^2})$ and $B = b/(\sqrt{a^2 + b^2})$ belong to a circle of radius 1 with the center at the origin, i.e., $A^2 + B^2 = 1$. Therefore, there is an angle ψ such that $\sin\psi = A$ and $\cos\psi = B$. For $A, B \geq 0$, the angle

ψ is determined using the equation $\psi = \tan^{-1}(a/b)$. The original equation takes the form

$$\sin\psi\cos x + \cos\psi\sin x = \frac{c}{\sqrt{a^2 + b^2}}$$

$$\Longleftrightarrow \sin(x + \psi) = \frac{c}{\sqrt{a^2 + b^2}}.$$

The resulting equation is easy to solve.

Remark 5.2. Similarly, we can show that there exists an angle ϕ such that $\cos\phi = A$ and $\sin\phi = B$. For $A, B \geq 0$, the angle ϕ is determined using the equation $\phi = \tan^{-1}(b/a)$. The original equation now takes the form

$$\cos(x - \phi) = \frac{c}{\sqrt{a^2 + b^2}}.$$

For $A, B \geq 0$, the angles ψ and ϕ are related by the ratio $\psi = \pi/2 - \phi$ (see Fig. 5.1).

Example 5.9. Solve the inequality

$$3\sin 2\pi x \geq \sqrt{2}\sin 4\pi x + 3\cos 2\pi x + \sqrt{32}.$$

Proof. Let's write the inequality in the following form:

$$3\sin 2\pi x - 3\cos 2\pi x \geq \sqrt{2}(\sin 4\pi x + 4).$$

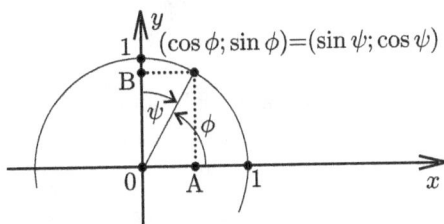

Fig. 5.1.   Trigonometric circle.

Transform the expression on the left by introducing an auxiliary angle:

$$3\sin 2\pi x - 3\cos 2\pi x = 3\sqrt{2}\left(\frac{1}{\sqrt{2}}\sin 2\pi x - \frac{1}{\sqrt{2}}\cos 2\pi x\right)$$
$$= 3\sqrt{2}\left(\cos(\pi/4)\cdot\sin 2\pi x - \sin(\pi/4)\cdot\cos 2\pi x\right)$$
$$= 3\sqrt{2}\sin(2\pi x - \pi/4).$$

Thus, the initial inequality takes the form

$$3\sqrt{2}\sin(2\pi x - \pi/4) \geq \sqrt{2}(\sin 4\pi x + 4).$$

We get

$$3\sqrt{2} \geq 3\sqrt{2}\sin(2\pi x - \pi/4) \geq \sqrt{2}(\sin 4\pi x + 4) \geq 3\sqrt{2}$$

since the functions $\sin x$ and $\cos x$ do not exceed one in the modulus. It follows from these inequalities that the initial inequality occurs only in the case of

$$\begin{cases} 3\sqrt{2}\sin(2\pi x - \pi/4) = 3\sqrt{2}, \\ \sqrt{2}(\sin 4\pi x + 4) = 3\sqrt{2}. \end{cases}$$

$$\Longleftrightarrow \begin{cases} \sin(2\pi x - \pi/4) = 1, \\ \sin 4\pi x = -1. \end{cases}$$

$$\Longleftrightarrow \begin{cases} 2\pi x - \pi/4 = \pi/2 + 2\pi m, \quad m \in \mathbb{Z}, \\ 4\pi x = -\pi/2 + 2\pi n, \quad n \in \mathbb{Z}. \end{cases}$$

$$\Longleftrightarrow \begin{cases} x = 3/8 + m, \quad m \in \mathbb{Z}, \\ x = -1/8 + n/2, \quad n \in \mathbb{Z}. \end{cases}$$

It remains to be noted that the set of solutions of the first equation in the system is completely included in the set of solutions of the second equation (for this, it is enough to put $n = 2l + 1$, $l \in \mathbb{Z}$.)

Answer: $3/8 + m$, $m \in \mathbb{Z}$. $\square$

Example 5.10. Find the domain of definition of the function

$$f(x) = \frac{1}{\sqrt{-6\sin^2 2x - 2\sin 2x \cos 2x + 8 - \sqrt{3}}}.$$

Proof. The domain of definition $f(x)$ is given by the inequality

$$-6\sin^2 2x - 2\sin 2x \cos 2x + 8 - \sqrt{3} > 0.$$

Transform the expression on the left by introducing an auxiliary angle:

$$-6\sin^2 2x - 2\sin 2x \cos 2x + 8 - \sqrt{3}$$

$$= -3(1 - \cos 4x) - \sin 4x + 8 - \sqrt{3}$$

$$= 3\cos 4x - \sin 4x + 5 - \sqrt{3}$$

$$= \sqrt{10}\left(\frac{3}{\sqrt{10}}\cos 4x - \frac{1}{\sqrt{10}}\sin 4x\right) + 5 - \sqrt{3}$$

$$= \sqrt{10}\left(\cos\alpha \cos 4x - \sin\alpha \sin 4x\right) + 5 - \sqrt{3}$$

$$= \sqrt{10}\cos(4x + \alpha) + 5 - \sqrt{3},$$

where $\alpha = \tan^{-1}(1/3)$. From the inequality

$$\sqrt{10}\cos(4x + \alpha) + 5 - \sqrt{3} \geq -\sqrt{10} + 5 - \sqrt{3},$$

if we prove that $-\sqrt{10} + 5 - \sqrt{3} > 0$, it follows that the domain of definition of the function $f(x)$ is the set of all real numbers. In fact,

$$5 - \sqrt{3} \bigvee \sqrt{10} \iff 28 - 10\sqrt{3} \bigvee 10$$

$$\iff 18 \bigvee 10\sqrt{3} \iff 9 \bigvee 5\sqrt{3}$$

$$\iff 81 \bigvee 75.$$

Thus, the inequality is proved: $-\sqrt{10} + 5 - \sqrt{3} > 0$.

Answer: $\mathbb{R}$. □

Example 5.11. Find all solutions of the equation

$$\frac{1}{\sqrt{2}} \sin^2 \left(x + \frac{\pi}{12} \right) + \sin 3x = \cos 3x - \sqrt{2}$$

on the segment $[-2\pi; 2\pi]$.

Proof. Transform the expression $\sin 3x - \cos 3x$ by introducing an auxiliary angle:

$$\sin 3x - \cos 3x = \sqrt{2} \left(\frac{1}{\sqrt{2}} \sin 3x - \frac{1}{\sqrt{2}} \cos 3x \right)$$

$$= \sqrt{2}(\cos(\pi/4) \sin 3x - \sin(\pi/4) \cos 3x) = \sqrt{2} \sin(3x - \pi/4).$$

Let us write the original equation in the following form:

$$\frac{1}{\sqrt{2}} \sin^2 \left(x + \frac{\pi}{12} \right) + \sqrt{2} \left(\sin(3x - \pi/4) + 1 \right) = 0.$$

Since both terms are non-negative,

$$\sin^2 \left(x + \frac{\pi}{12} \right) \geq 0, \quad \sin(3x - \pi/4) + 1 \geq 0,$$

then the left side of the equation can be equal to zero only if both terms are equal to zero, i.e.,

$$\begin{cases} \sin^2(x + \pi/12) = 0, \\ \sin(3x - \pi/4) + 1 = 0. \end{cases} \iff \begin{cases} \sin(x + \pi/12) = 0, \\ \sin(3x - \pi/4) = -1. \end{cases}$$

$$\iff \begin{cases} x = -\pi/12 + \pi m, \quad m \in \mathbb{Z}, \\ 3x - \pi/4 = -\pi/2 + 2\pi n, \quad n \in \mathbb{Z}. \end{cases}$$

$$\iff \begin{cases} x = -\pi/12 + \pi m, \quad m \in \mathbb{Z}, \\ x = -\pi/12 + 2\pi n/3, \quad n \in \mathbb{Z}. \end{cases}$$

Let us plot solutions of the first equation of the system on a trigonometric circle (see Fig. 5.2) and do the same with the second equation (see Fig. 5.3).

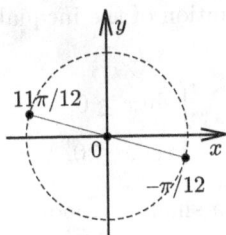

Fig. 5.2. Solution of $\sin\left(x + \frac{\pi}{12}\right) = 0$.

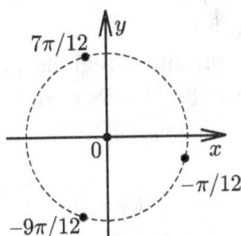

Fig. 5.3. Solution of $\sin\left(3x - \frac{\pi}{4}\right) = -1$.

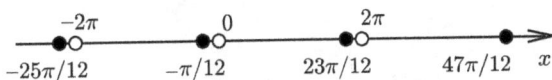

Fig. 5.4.   Roots on the segment $[0; 2\pi]$.

From this, by intersecting the obtained solutions, we get

$$x = -\pi/12 + 2\pi n, \quad n \in \mathbb{Z}.$$

It remains to select the roots located on the segment $[0; 2\pi]$ (see Fig. 5.4). To do this, it is enough to solve the inequality

$$- 2\pi \le -\pi/12 + 2\pi n \le 2\pi$$
$$\Longleftrightarrow -23\pi/12 \le 2\pi n \le 25\pi/12$$
$$\Longleftrightarrow -23/24 \le n \le 25/24 \Longleftrightarrow n = 0, \quad n = 1.$$

Answer: $-\pi/12$, $2\pi - \pi/12$. □

Example 5.12. Solve the inequality

$$\sqrt{\sin x} > \sqrt{-\cos x}.$$

Proof. The domain of definition of the inequality is determined using the conditions

$$\begin{cases} \sin x \geq 0, \\ \cos x \leq 0. \end{cases}$$

For clarity, we depict it on a single trigonometric circle (Fig. 5.5). On the domain of definition, the initial inequality is equivalent to the inequality $\sin x > -\cos x$, or $\sin x + \cos x > 0$, or $\sqrt{2}\sin\left(x + \dfrac{\pi}{4}\right) > 0$. We get $2\pi n < x + \pi/4 < 2\pi n + \pi$ or $2\pi n - \pi/4 < x < 2\pi n + 3\pi/4$. On the circle, this area is depicted as shown in Fig. 5.6.

The intersection of the obtained regions (see Fig. 5.7), given by the inequalities $\pi/2 + 2\pi n \leq x < 3\pi/4 + 2\pi n$ and $n \in \mathbb{Z}$, gives the answer to the problem. □

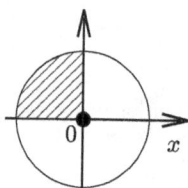

Fig. 5.5. $\sin x \geq 0$ and $\cos x \geq 0$.

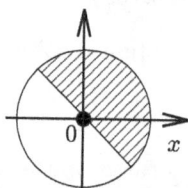

Fig. 5.6. $\sin x > -\cos x$.

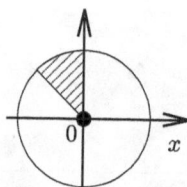

Fig. 5.7.   Intersection.

Example 5.13. Find all the numbers x on the interval $[-\pi, \pi]$ satisfying the inequalities

$$(4 + \sqrt{3}) \sin x + 2\sqrt{3} + 1 \leq \cos 2x \leq 5 \cos x - 3.$$

Proof. We solve each inequality separately. Replace $\cos 2x$ by the double angle formula through $1 - 2\sin^2 x$ in the first inequality and through $2\cos^2 x - 1$ in the second:

$$\begin{cases} (4 + \sqrt{3}) \sin x + 2\sqrt{3} + 1 \leq \cos 2x, \\ \cos 2x \leq 5 \cos x - 3. \end{cases}$$

$$\Longleftrightarrow \begin{cases} 2\sin^2 x + (4 + \sqrt{3}) \sin x + 2\sqrt{3} \leq 0, \\ 2\cos^2 x - 5 \cos x + 3 \leq 0. \end{cases}$$

$$\Longleftrightarrow \begin{cases} 2(\sin x + 2)(\sin x + \sqrt{3}/2) \leq 0, \\ 2(\cos x - 1/2)(\cos x - 2) \leq 0. \end{cases}$$

$$\Longleftrightarrow \begin{cases} -2 \leq \sin x \leq -\sqrt{3}/2, \\ 1/2 \leq \cos x \leq 2. \end{cases}$$

In Fig. 5.8, we depict the sets of solutions to these inequalities. They intersect only at one point, namely $x = -\pi/3 + 2\pi n$, $n \in \mathbb{Z}$. The segment $[-\pi; \pi]$ has only a single value $x = -\pi/3$.

Answer: $-\pi/3$. $\qquad\qquad\qquad\qquad\qquad\qquad\qquad\qquad\qquad\square$

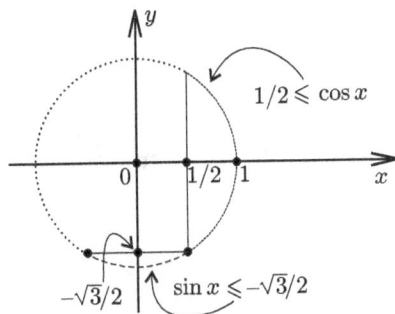

Fig. 5.8. Set of solution.

5.3   Tasks Requiring Additional Research

Here, we consider problems that require either root selection or in which it is important to use the limitations of trigonometric functions. It is important to remember that the modulus of the sine and cosine functions is no more than 1. Sometimes, this consideration plays a key role in solving the problem. Let us look at a few examples.

Example 5.14. Solve the equation

$$\sqrt{\sin 2x} = \sqrt{3(\cos x + \sin x) - 3}.$$

Proof. We use the identity

$$(\cos x + \sin x)^2 = \cos^2 x + \sin^2 x + 2\cos x \sin x = 1 + \sin 2x.$$

Then, $\sin 2x = (\cos x + \sin x)^2 - 1$, and when replacing the variable $t = \cos x + \sin x$, the original equation takes the form

$$\sqrt{t^2 - 1} = \sqrt{3t - 3}.$$

The domain of definition of the resulting equation is determined by the inequalities

$$\begin{cases} t^2 - 1 \geq 0, \\ t - 1 \geq 0. \end{cases} \iff t \geq 1.$$

Squaring both sides of the equation, we come to the equation

$$t^2 - 1 = 3t - 3,$$

or

$$t^2 - 3t + 2 = 0,$$

whose roots $t_1 = 1$ and $t_2 = 2$ both belong to the domain of definition. It remains to solve the equations

$$\cos x + \sin x = 1 \quad \text{and} \quad \cos x + \sin x = 2.$$

After the transformation

$$\cos x + \sin x = \sqrt{2}\sin\left(x + \frac{\pi}{4}\right),$$

we get

$$\sin\left(x + \frac{\pi}{4}\right) = \frac{1}{\sqrt{2}} \quad \text{and} \quad \sin\left(x + \frac{\pi}{4}\right) = \sqrt{2}.$$

The second of them has no solutions, and from the first, we find

$$x + \frac{\pi}{4} = (-1)^n \frac{\pi}{4} + \pi n, \quad n \in \mathbb{Z},$$

or

$$x = -\frac{\pi}{4} + (-1)^n \frac{\pi}{4} + \pi n, \quad n \in \mathbb{Z}.$$

This set can be represented in another form:

$$x = 2\pi k, \quad k \in \mathbb{Z} \quad \text{and} \quad x = \frac{\pi}{2} + 2\pi l, \quad l \in \mathbb{Z}.$$

Answer: $x = 2\pi k$, $x = \pi/2 + 2\pi l$, $k, l \in \mathbb{Z}$. $\qquad\square$

Example 5.15. Solve the equation

$$\tan x + \tan 2x + \tan x \tan 2x \tan 3x = \tan 3x + \tan 4x.$$

Proof. Let us find the domain of definition of this equation:

$$\begin{cases} \cos x \neq 0, \\ \cos 2x \neq 0, \\ \cos 3x \neq 0, \\ \cos 4x \neq 0. \end{cases}$$

Now, according to the well-known formula

$$\tan(x \pm y) = \frac{\tan x \pm \tan y}{1 \mp \tan x \tan y}, \quad x, y, x \pm y \neq \pi(2n+1)/2, \quad n \in \mathbb{Z}.$$

We have

$$\tan x = \tan(3x - 2x) = \frac{\tan 3x - \tan 2x}{1 + \tan 2x \tan 3x}$$

$$\Longleftrightarrow \tan x + \tan 2x + \tan x \tan 2x \tan 3x = \tan 3x.$$

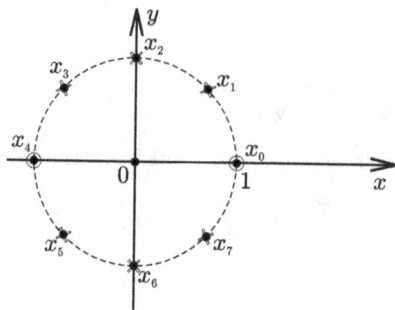

Fig. 5.9.   Roots of equation.

Consequently, the original equation is transformed to the form

$$\tan 4x = 0$$

$$\Longleftrightarrow x = \pi n/4, \quad n \in \mathbb{Z}.$$

To check the fulfilment of the conditions defining acceptable values, consider the points $x_n = \pi n/4$. They are depicted on a trigonometric circle, as shown in Fig. 5.9:

- For $n = 0$, we have $\cos 0 = \cos 2 \cdot 0 = \cos 3 \cdot 0 = \cos 4 \cdot 0 = 1 \neq 0$.
- For $n = 1$, we have $\cos(2 \cdot \pi/4) = 0$ (the condition $\cos 2x \neq 0$ is not satisfied).
- For $n = 2$, we have $\cos(3 \cdot 2\pi/4) = 0$ (the condition $\cos 3x \neq 0$ is not satisfied).
- For $n = 3$, we have $\cos(2 \cdot 3\pi/4) = 0$ (the condition $\cos 2x \neq 0$ is not satisfied).
- For $n = 4$, we have $-\cos \pi = \cos(2 \cdot \pi) = -\cos(3 \cdot \pi) = \cos(4 \cdot \pi) = 1 \neq 0$.
- For $n = 5$, we have $\cos(2 \cdot 5\pi/4) = 0$ (the condition $\cos 2x \neq 0$ is not satisfied).
- For $n = 6$, we have $\cos(3 \cdot 6\pi/4) = 0$ (the condition $\cos 3x \neq 0$ is not satisfied).
- For $n = 7$, we have $\cos(2 \cdot 7\pi/4) = 0$ (the condition $\cos 2x \neq 0$ is not satisfied).

As a result of the check, we get that the points satisfy the conditions of the problem (see Fig. 5.10), which can be set using the formula $x = \pi n$.

Answer: πn, $n \in \mathbb{Z}$.   □

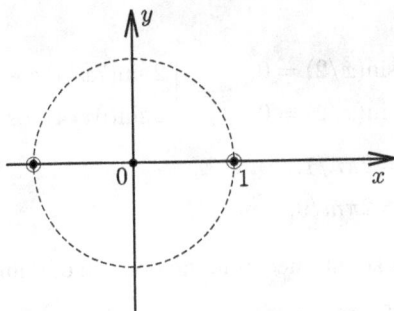

Fig. 5.10. Roots after selection.

Example 5.16. Solve the equation

$$(63\cos^2(x/2) - \sin^2(x/2))\cos^2 x = \tan^2(2x) + \sin^2 x.$$

Proof. Let us find the domain of definition of this equation:

$$\cos 2x \neq 0 \Longleftrightarrow 2x \neq (\pi/2) + \pi k, \quad k \in \mathbb{Z}$$
$$\Longleftrightarrow x \neq (\pi/4) + \pi k/2, \quad k \in \mathbb{Z}.$$

We transform the original equation to the following form:

$$(64\cos^2(x/2) - 1)\cos^2 x = \tan^2(2x) + \sin^2 x$$
$$\Longleftrightarrow 64\cos^2(x/2)\cos^2 x = \tan^2(2x) + 1$$
$$\Longleftrightarrow 64\cos^2(x/2)\cos^2 x \cos^2 2x = 1.$$

First of all, we check by substitution that $x = 2\pi l$ and $l \in \mathbb{Z}$ are not solutions of the equation. Multiplying by $\sin^2(x/2)$, $x \neq 2\pi l$, $l \in \mathbb{Z}$, we get

$$16 \cdot \left(4\sin^2(x/2)\cos^2(x/2)\right) \cdot \cos^2 x \cos^2 2x = \sin^2(x/2)$$
$$\Longleftrightarrow 16\sin^2 x \cos^2 x \cos^2 2x = \sin^2(x/2)$$
$$\Longleftrightarrow 4\sin^2 2x \cos^2 2x = \sin^2(x/2)$$
$$\Longleftrightarrow \sin^2 4x = \sin^2(x/2).$$

So,

$$\left[\begin{array}{l} \sin(4x) - \sin(x/2) = 0, \\ \sin(4x) + \sin(x/2) = 0. \end{array}\right. \iff \left[\begin{array}{l} 2\sin(7x/4)\cos(9x/4) = 0, \\ 2\sin(9x/4)\cos(7x/4) = 0. \end{array}\right.$$

$$\iff \left[\begin{array}{ll} x = 2\pi n/7, & n \in \mathbb{Z}, \\ x = 2\pi m/9, & m \in \mathbb{Z}. \end{array}\right.$$

Now, it remains to take into account the domain of definition. The equality

$$\frac{2\pi n}{7} = \frac{\pi}{4} + 2\pi k \iff 4n = \frac{7}{2} + 7k$$

is impossible since the left part of the equation $4n = 7/2 + 7k$ is an integer $(n, k \in \mathbb{Z})$ and the right part is not. Similarly, it is verified that the equality $2\pi m/9 = (\pi/4) + \pi k/2$, $k, m \in \mathbb{Z}$ is impossible. Thus, the obtained set of points belong to the domain of definition. It remains to discard the extra roots (which arose when multiplying the equation by the function $\sin^2(x/2)$):

$$x \neq 2\pi l, \quad l \in \mathbb{Z}.$$

With the help of equivalent relations

$$\frac{2\pi n}{7} \neq 2\pi l \iff n \neq 7l, \quad l \in \mathbb{Z},$$

$$\frac{2\pi m}{9} \neq 2\pi l \iff m \neq 9l, \quad l \in \mathbb{Z},$$

we arrive at the answer.

Answer: $2\pi n/7$, $2\pi m/9$, $n, m \in \mathbb{Z}$, $n \neq 7l$, $m \neq 9l$, $l \in \mathbb{Z}$.   □

Remark 5.3. In solving root selection problems, the following simple theorem is useful (although, of course, you can do without it).

Theorem. Let a and b be coprime integers, $c \in \mathbb{Z}$. If $(x_0; y_0)$ is an integer solution of the equation $ax + by + c = 0$, then the solutions

$$x = x_0 + bn, \quad y = y_0 - an, \quad n \in \mathbb{Z},$$

make up the set of all solutions of the equation $ax + by + c = 0$ in integer numbers.

Example 5.17. Solve the equation

$$\cos 6x - 3\cos 5x + \cos 4x - 4\cos x + 5 = 0.$$

Proof. Using the cosine sum formula $\cos\alpha + \cos\beta = 2\cos((\alpha+\beta)/2)\cos((\alpha+\beta)/2)$, we get that the original equation is equivalent to the equation

$$\cos 6x - 3\cos 5x + \cos 4x - 4\cos x + 5 = 0$$
$$\Longleftrightarrow 2\cos 5x\cos x - 3\cos 5x - 4\cos x + 5 = 0$$
$$\Longleftrightarrow (2\cos x - 3)(\cos 5x - 2) = 1.$$

However, since $|\cos y| \leq 1$, then $2\cos x - 3 \leq -1$, $\cos 5x - 2 \leq -1$, and hence

$$(2\cos x - 3)(\cos 5x - 2) \geq 1.$$

The sign of equality in the latter inequality is possible only in the case of

$$\begin{cases} 2\cos x - 3 = -1, \\ \cos 5x - 2 = -1. \end{cases} \Longleftrightarrow \begin{cases} \cos x = 1, \\ \cos 5x = 1. \end{cases}$$
$$\Longleftrightarrow \cos x = 1 \Longleftrightarrow x = \pi/2 + 2\pi n, \quad n \in \mathbb{N}.$$

Answer: $\pi/2 + 2\pi n$, $n \in \mathbb{N}$. □

Example 5.18. Solve the equation

$$\cos^4 x = \frac{1}{4}\cos 2x + \frac{1}{2}\cos^2 x\cos 8x.$$

Proof. First, we use the well-known identity $\cos 2x = 2\cos^2 x - 1$ and proceed to the equation

$$\cos^4 x = \frac{1}{2}\cos^2 x - \frac{1}{4} + \frac{1}{2}\cos^2 x\cos 8x,$$

or

$$\cos^4 x - \frac{1}{2}\cos^2 x + \frac{1}{4} = \frac{1}{2}\cos^2 x\cos 8x.$$

Subtracting from both parts $(\cos^2 x)/2$ in order to get a full square on the left part, we get the equation

$$\cos^4 x - \cos^2 x + \frac{1}{4} = \frac{1}{2} \cos^2 x \cos 8x - \frac{1}{2} \cos^2 x,$$

or

$$\left(\cos^2 x - \frac{1}{2}\right)^2 = \frac{1}{2} \cos^2 x(\cos 8x - 1).$$

The left-hand side of this equation is always non-negative, and the right-hand side is non-positive, since $\cos 8x - 1 \le 0$, for all x, and $\cos^2 x \ge 0$, for all x. Therefore, equality holds if and only if

$$\begin{cases} \cos 8x = 1, \\ \cos^2 x = \frac{1}{2}. \end{cases} \iff \begin{cases} \cos 8x = 1, \\ 2\cos^2 x - 1 = 0. \end{cases} \iff \begin{cases} \cos 8x = 1, \\ \cos 2x = 0. \end{cases}$$

Solving the second equation of the system, we find

$$x = \frac{\pi}{4} + \frac{\pi k}{2}, \quad k \in \mathbb{Z}.$$

But at the same time, $8x = 8\pi/4 + 8\pi k/2 = 2\pi + 4\pi k$, $k \in \mathbb{Z}$, and $\cos 8x = 1$, so that the first equality of the system is fulfilled.

Answer: $x = \pi/4 + \pi k/2$, $k \in \mathbb{Z}$.   □

Example 5.19. Solve the equation

$$\sin x(\cos 2x + \cos 6x) + \cos^2 x = 2.$$

Proof. Let us solve the example in two ways:

1. Replace $\cos^2 x$ by $1 - \sin^2 x$, and select the full squares in the equation:

$$\sin x(\cos 2x + \cos 6x) + \cos^2 x = 2$$

$$\iff \sin^2 x - \sin x(\cos 2x + \cos 6x) + 1 = 0$$

$$\iff \left(\sin x - \frac{\cos 2x + \cos 6x}{2}\right)^2 + 1 - \left(\frac{\cos 2x + \cos 6x}{2}\right)^2 = 0.$$

However, since the value of the expression $|\cos 2x + \cos 6x|$ does not exceed 2,

$$\begin{cases} 1 - ((\cos 2x + \cos 6x)/2)^2 \geq 0, \\ (\sin x - (\cos 2x + \cos 6x)/2)^2 \geq 0. \end{cases}$$

We have obtained that the sum of two non-negative terms is zero. Therefore, each of them is equal to zero:

$$\begin{cases} 1 - ((\cos 2x + \cos 6x)/2)^2 = 0, \\ (\sin x - (\cos 2x + \cos 6x)/2)^2 = 0. \end{cases}$$

$$\Longleftrightarrow \begin{cases} |\cos 2x + \cos 6x| = 2, \\ \cos 2x + \cos 6x = 2\sin x. \end{cases}$$

$$\Longleftrightarrow \begin{cases} \cos 2x + \cos 6x = \pm 2, \\ \sin x = \pm 1. \end{cases}$$

$$\Longleftrightarrow \left[\begin{array}{l} \begin{cases} \cos 2x + \cos 6x = 2, \\ \sin x = 1. \end{cases} \\ \begin{cases} \cos 2x + \cos 6x = -2, \\ \sin x = -1. \end{cases} \end{array} \right. \Longleftrightarrow \left[\begin{array}{l} \begin{cases} \cos 2x = 1, \\ \cos 6x = 1, \\ \sin x = 1. \end{cases} \\ \begin{cases} \cos 2x = -1, \\ \cos 6x = -1, \\ \sin x = -1. \end{cases} \end{array} \right.$$

Let us find the solutions to the first system in the aggregate. Solving the equation $\sin x = 1$, we find $x = \pi/2 + 2\pi n$, $n \in \mathbb{Z}$, but these solutions do not satisfy the equation $\cos 2x = 1$. Therefore, there are no solutions to the first system in the aggregate.

Similarly, we find solutions to the second system in the aggregate. Solving the equation $\sin x = -1$, we find $x = -\pi/2 + 2\pi n$, $n \in \mathbb{Z}$, and by substitution, we make sure that the found values of x satisfy the equations $\cos 2x = -1$ and $\cos 6x = -1$. Therefore, the solution of the aggregate will be the set $x = -\pi/2 + 2\pi n$, $n \in \mathbb{Z}$.

2. We demonstrate the solution using the discriminant method. The equation

$$\sin^2 x - \sin x(\cos 2x + \cos 6x) + 1 = 0$$

could be considered a square relative to the variable $\sin x$. Find the discriminant of this equation:

$$D = (\cos 2x + \cos 6x)^2 - 4 \leq 0.$$

In order for solutions to exist, the condition $D = 0$ is necessary, which is equivalent to the system

$$\begin{cases} |\cos 2x + \cos 6x| = 2, \\ \cos 2x + \cos 6x = 2 \sin x. \end{cases}$$

We solve it as we did in the first method.

Answer: $x = -\pi/2 + 2\pi n$, $n \in \mathbb{Z}$. $\square$

5.4   Trigonometric Equations and Inequalities with the Parameter

Problems often use the limitation of the functions $\sin x$ and $\cos x$, as well as the auxiliary argument method.

Example 5.20. For each value of a, solve the equation

$$4 \cos x \sin a + 2 \sin x \cos a - 3 \cos a = 2\sqrt{7}.$$

Proof. The equation has the following form:

$$A(a) \cos x + B(a) \sin x = C(a),$$

where

$$A(a) = 4 \sin a,$$
$$B(a) = 2 \cos a,$$
$$C(a) = 2\sqrt{7} + 3 \cos a.$$

We transform the equation in the way indicated above. We get

$$\sqrt{A^2(a) + B^2(a)}(\sin(x + \varphi(a))) = C(a).$$

Calculate

$$\sqrt{A^2(a) + B^2(a)} = \sqrt{16 \sin^2 a + 4 \cos^2 a} = \sqrt{12 \sin^2 a + 4} \geq 2 > 0.$$

Therefore, the equation is equivalent to

$$\sin(x + \varphi(a)) = \frac{C(a)}{\sqrt{A^2(a) + B^2(a)}}.$$

This equation will admit a solution if and only if

$$\left|\frac{C(a)}{\sqrt{A^2(a) + B^2(a)}}\right| \le 1 \iff |C(a)| \le \sqrt{A^2(a) + B^2(a)}$$

$$\iff (2\sqrt{7} + 3\cos a)^2 \le 16\sin^2 a + 4\cos^2 a$$

$$\iff 28 + 12\sqrt{7}\cos a + 9\cos^2 a \le 16(1 - \cos^2 a) + 4\cos^2 a$$

$$\iff 12 + 12\sqrt{7}\cos a + 21\cos^2 a \le 0$$

$$\iff 4 + 4\sqrt{7}\cos a + 7\cos^2 a \le 0$$

$$\iff (2 + \sqrt{7}\cos a)^2 \le 0.$$

This means that the equation has a solution only when $\cos a = -2/\sqrt{7}$. Two cases are possible:

$$\text{I.} \begin{cases} \cos a = -2/\sqrt{7}, \\ \sin a = \sqrt{3/7}. \end{cases} \qquad \text{II.} \begin{cases} \cos a = -2/\sqrt{7}, \\ \sin a = -\sqrt{3/7}. \end{cases}$$

Substitute the found value into the original equation.

• In the first case ($\cos a = -2/\sqrt{7}$, $\sin a = \sqrt{3/7}$),

$$4\cos x \cdot \sqrt{\frac{3}{7}} + 2\sin x \cdot \left(-\frac{2}{\sqrt{7}}\right) + \frac{6}{\sqrt{7}} = 2\sqrt{7}$$

$$\iff 4\sqrt{3}\cos x - 4\sin x + 6 = 14 \iff \sqrt{3}\cos x - \sin x = 2$$

$$\iff 2\cos\left(x + \frac{\pi}{6}\right) = 2 \iff \cos\left(x + \frac{\pi}{6}\right) = 1,$$

or $x = -\pi/6 + 2\pi k$, $k \in \mathbb{Z}$.

- In the second case $(\cos a = -2/\sqrt{7},\ \sin a = -\sqrt{3/7})$,

$$4\cos x \cdot \left(-\sqrt{\frac{3}{7}}\right) + 2\sin x \cdot \left(-\frac{2}{\sqrt{7}}\right) + \frac{6}{\sqrt{7}} = 2\sqrt{7}$$

$$\Longleftrightarrow -4\sqrt{3}\cos x - 4\sin x + 6 = 14$$

$$\Longleftrightarrow -\sqrt{3}\cos x - \sin x = 2 \Longleftrightarrow 2\cos\left(x - \frac{\pi}{6}\right) = -2$$

$$\Longleftrightarrow \cos\left(x - \frac{\pi}{6}\right) = -1,$$

or $x = \pi/6 - \pi + 2\pi k = -5\pi/6 + 2\pi n,\ n \in \mathbb{Z}$.

Answer: If $a = \arccos(-2/\sqrt{7}) + 2\pi l,\ l \in \mathbb{Z}$, then $x = -\pi/6 + 2\pi k,\ k \in \mathbb{Z}$; if $a = -\arccos(-2/\sqrt{7}) + 2\pi m,\ m \in \mathbb{Z}$, then $x = -5\pi/6 + 2\pi n,\ n \in \mathbb{Z}$; there are no solutions for other values of the parameter a. □

Example 5.21. Find all the values of the parameter α for each of which the system of equations

$$\begin{cases} \sin x = \cos(x\sqrt{6 - 2a^2}), \\ \cos x = (a - 2/3)\sin(x\sqrt{6 - 2a^2}) \end{cases}$$

has exactly one solution on the segment $[0; 2\pi]$.

Proof. Let us rewrite the system in the following form:

$$\begin{cases} \sin x = \sin(\pi/2 - x\sqrt{6 - 2a^2}), \\ \cos x = (a - 2/3)\cos(\pi/2 - x\sqrt{6 - 2a^2}), \end{cases}$$

and we introduce the notation

$$\alpha = \alpha(x, a) = \pi/2 - x\sqrt{6 - 2a^2}.$$

Then, the original system takes the form

$$\begin{cases} \sin x = \sin \alpha, \\ \cos x = (a - 2/3)\cos \alpha. \end{cases}$$

The equality $\sin x = \sin \alpha$ means that the corresponding cosines can differ only by a sign, i.e.,

$$\begin{cases} \sin x = \sin \alpha, \\ \cos x = (a - 2/3)\cos\alpha. \end{cases} \Longleftrightarrow \left[\begin{array}{l} \begin{cases} \sin x = \sin \alpha, \\ \cos x = \cos \alpha, \\ \cos x = (a - 2/3)\cos\alpha; \end{cases} \\ \begin{cases} \sin x = \sin \alpha, \\ \cos x = -\cos \alpha, \\ \cos x = (a - 2/3)\cos\alpha. \end{cases} \end{array}\right.$$

$$\Longleftrightarrow \left[\begin{array}{l} \begin{cases} \sin x = \sin \alpha, \\ \cos x = \cos \alpha, \\ \cos x \cdot (5/3 - a) = 0; \end{cases} \\ \begin{cases} \sin x = \sin \alpha, \\ \cos x = -\cos \alpha, \\ \cos x \cdot (1/3 + a) = 0. \end{cases} \end{array}\right. \Longleftrightarrow \left[\begin{array}{l} x = \pi/2, \\ x = 3\pi/2, \\ a = 5/3, \\ a = -1/3. \end{array}\right.$$

(solutions are considered only from the segment $[0; 2\pi]$). Let us analyze all four cases.

1. Let $a = -1/3$. Then, the system takes the form

$$\begin{cases} \sin x = \sin \alpha, \\ \cos x = -\cos \alpha. \end{cases} \Longleftrightarrow \alpha = (\pi - x) + 2\pi n, \quad n \in \mathbb{Z}.$$

But $\alpha = \pi/2 - x \cdot 2\sqrt{13}/3$. Hence,

$$\alpha = (\pi - x) + 2\pi n \Longleftrightarrow x = x_n = \frac{-\pi/2 + 2\pi n}{2\sqrt{13}/3 - 1}, \quad n \in \mathbb{Z}.$$

We show that, among the found values x_n, only x_1 belongs to the segment $[0; 2\pi]$. First, we prove that the number $2\sqrt{13}/3 - 1$ belongs to the interval $(1; 5/3)$. Indeed,

$$2\sqrt{13}/3 - 1 > 2 \cdot 3/3 - 1 = 1,$$
$$2\sqrt{13}/3 - 1 < 2 \cdot 4/3 - 1 = 8/3 - 1 = 5/3.$$

We carry out the selection of roots:

$$x_n = \frac{-\pi/2 + 2\pi n}{2\sqrt{13}/3 - 1} \le \frac{-\pi/2}{2\sqrt{13}/3 - 1} = x_0 < 0, \quad n \le 0,$$

$$x_1 = \frac{3\pi/2}{2\sqrt{13}/3 - 1} \in (0; 3\pi/2),$$

$$x_n = \frac{-\pi/2 + 2\pi n}{2\sqrt{13}/3 - 1} \ge \frac{7\pi/2}{2\sqrt{13}/3 - 1} = x_2 > \frac{7\pi/2}{5/3} = \frac{21\pi}{10} > 2\pi, \quad n \ge 2.$$

Conclusion: For $a = -1/3$, the original system of equations really has a unique solution,

$$x = \frac{3\pi/2}{2\sqrt{13}/3 - 1},$$

on the segment $[0; 2\pi]$.

2. Let $a = 5/3$. Then, the system takes the form

$$\begin{cases} \sin x = \sin \alpha, \\ \cos x = \cos \alpha. \end{cases} \iff \alpha = x + 2\pi n, \quad n \in \mathbb{Z}.$$

But $\alpha = \pi/2 - x \cdot 2/3$. Hence,

$$\alpha = x + 2\pi n \iff x = x_n = 3\pi/10 + 6\pi/5n, \quad n \in \mathbb{Z}.$$

Among the found values of x_n, to the segment $[0; 2\pi]$ belong $x_0 = 3\pi/10$ and $x_1 = 15\pi/10 = 3\pi/2$.

Conclusion: For $a = 5/3$, the initial system of equations has two solutions on the segment $[0; 2\pi]$. That is, the conditions of the problem are not met for $a = 5/3$.

3. Let $x = \pi/2$. Then, the system takes the form

$$\begin{cases} 1 = \sin \alpha, \\ 0 = (a - 2/3) \cos \alpha, \end{cases}$$

where $\alpha = \pi/2 \cdot (1 - \sqrt{6 - 2a^2})$. From the first equation, we find

$$\alpha = \pi/2 + 2\pi n \iff 1 - \sqrt{6 - 2a^2} = 1 + 4n$$

$$\iff \sqrt{6 - 2a^2} = -4n, \quad n \in \mathbb{Z}.$$

However, since $\sqrt{6-2a^2} \in [0; \sqrt{6}]$, the solutions to the equation $\sqrt{6-2a^2} = -4n$, $n \in \mathbb{Z}$ are possible only when $n = 0$, i.e.,

$$\sqrt{6-2a^2} = 0 \Longleftrightarrow a = \pm\sqrt{3}.$$

With $a = \pm\sqrt{3}$, taking into account $a - 2/3 \neq 0$, the original system is equivalent to

$$\begin{cases} \sin x = 1, \\ \cos x = 0. \end{cases}$$

Therefore, the system has a unique solution, $x = \pi/2$, on the segment $[0; 2\pi]$.

Conclusion: For $a = \pm\sqrt{3}$ the original system of equations really has a unique solution: $x = \pi/2$.

4. Let $x = 3\pi/2$. Then, the system takes the form

$$\begin{cases} -1 = \sin\alpha, \\ 0 = (a - 2/3)\cos\alpha, \end{cases}$$

where $\alpha = \pi/2 \cdot (1 - 3\sqrt{6-2a^2})$. From the first equation, we find

$$\alpha = 3\pi/2 + 2\pi n \Longleftrightarrow 1 - 3\sqrt{6-2a^2} = 3 + 4n$$

$$\Longleftrightarrow \sqrt{6-2a^2} = -2/3 - 4n/3, \quad n \in \mathbb{Z}.$$

However, since $\sqrt{6-2a^2} \in [0; \sqrt{6}]$, then the solutions of the equation $\sqrt{6-2a^2} = -2/3 - 4n/3$, $n \in \mathbb{Z}$, are possible only if $n = -1, -2$, i.e.,

$$\sqrt{6-2a^2} = 2/3, 2 \Longleftrightarrow a = \pm 5/3, \pm 1.$$

The case $a = 5/3$ is parsed in point II, and this value does not fit. Note that for $a = -5/3, \pm 1$, the inequality $(a - 2/3) \neq \pm 1$ is valid. Hence, the system

$$\begin{cases} \sin x = \sin\alpha, \\ \cos x = (a - 2/3)\cos\alpha, \end{cases}$$

can only have the solutions $x = \pi/2, 3\pi/2$ (since it follows from the equality $\sin x = \sin\alpha$ that $|\cos x| = |\cos\alpha|$). The value of $x = \pi/2$ is possible only if $a = \pm\sqrt{3}$ (see point III). Therefore, $x = 3\pi/2$ remains.

However, as was shown above, $x = 3\pi/2$ is the solution of the system and, therefore, the only one because there are no other solutions.

Conclusion: For $a = -5/3, \pm 1$, the original system of equations really has a unique solution: $x = 3\pi/2$.

Answer: $a \in \{-1/3; -5/3; \pm 1; \pm \sqrt{3}\}$. $\qquad\qquad\qquad\qquad\qquad$ □

5.4.1 *Problems with the study of multiple solutions*

Example 5.22. Find the condition under which the distance between any two adjacent roots of the equation $p_3(\sin x) = 0$, where $p_3(t) = a_3 t^3 + a_2 t^2 + a_1 t + a_0$, does not exceed $\pi/3$.

Proof. Since the polynomial $p_3(t)$ has no more than three roots, the original equation has only one possibility $t_1 = 0$, $t_{2,3} = \pm\sqrt{3}/2$, where t_1, t_2, and t_3 are the roots of the equation $p_3(t) = 0$ (see Fig. 5.11). Therefore, $p_3(t) = a(t^2 - 3/4)t$, i.e., $a_3 = a$, $a_2 = 0$, $a_1 = -3a/4$, and $a_0 = 0$, where $a \in \mathbb{R}\backslash\{0\}$.

Answer: $a_3 = a$, $a_2 = 0$, $a_1 = -3a/4$, $a_0 = 0$, where $a \in \mathbb{R}\backslash\{0\}$. □

Example 5.23. Find all the values of a at which, for any root of the equation

$$3\cos\alpha\sin x + \sin\alpha\sin 3x = 2\sin 2\alpha\cos 2x - \sin 3x + \cos 3\alpha,$$

there is another root at a distance of no more than $\pi/3$ from it.

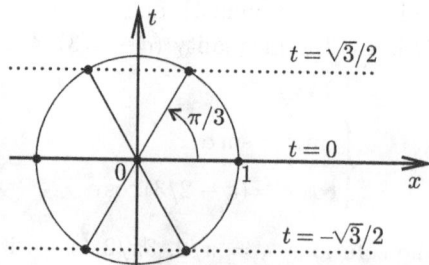

Fig. 5.11. Roots.

Proof. The original equation is equivalent to the following:

$$(1 + \sin\alpha)(3\sin x - 4\sin^3 x) - 2\sin 2\alpha(1 - 2\sin^2 x)$$
$$+ 3\cos\alpha \sin x - \cos 3\alpha = 0.$$

This equation is an equation of the third degree with respect to $\sin x$, and according to Example 5.22, the condition of the problem is equivalent to the fact that the original equation has the solutions $\sin x = 0$ and $\sin x = \pm\sqrt{3}/2$. Due to the periodicity of the functions, it is enough to substitute $x = 0$ and $x = \pm\pi/3$ into the original equation. Substituting these values, we find

$$x = 0 \Longrightarrow 2\sin 2\alpha + \cos 3\alpha = 0;$$
$$x = \pi/3 \Longrightarrow 3\cos\alpha \cdot (\sqrt{3}/2) = 2\sin 2\alpha \cdot (-1/2) + \cos 3\alpha;$$
$$x = -\pi/3 \Longrightarrow -3\cos\alpha \cdot (\sqrt{3}/2) = 2\sin 2\alpha \cdot (-1/2) + \cos 3\alpha.$$

The last two equations imply the equality $\cos\alpha = 0$, but then $\cos 3\alpha = 0$. It remains to be noted that from these three equations, it follows that $\sin 2\alpha$ must also be zero; however, this condition is fulfilled due to the fact that $\cos\alpha = 0$.

Answer: $\alpha = \pi/2 + \pi n$, $n \in \mathbb{Z}$. $\qquad\square$

Example 5.24. Find all the values of a and b for each of which the equation $p_2(\sin x) = 0$, where $p_2(t) = (t - a)(t - b)$, has solutions and all its positive solutions form an arithmetic progression.

Proof. We depict all possible cases on trigonometric circles (see Figs. 5.12–5.18).

1. The case in Fig. 5.12 is possible when $a = 1$, $b \in (-\infty; -1) \cup [1; +\infty)$ or $b = 1$, $a \in (-\infty; -1) \cup [1; +\infty)$.
2. The case in Fig. 5.13 is possible when $a = -1$, $b \in (-\infty; -1] \cup (1; +\infty)$ or $b = -1$, $a \in (-\infty; -1] \cup (1; +\infty)$.
3. The case in Fig. 5.14 is possible when $a = -1$, $b = 1$ or $b = -1$, $a = 1$.
4. The case in Fig. 5.15 is possible when $a = 0$, $b \in (-\infty; -1) \cup \{0\} \cup (1; +\infty)$ or $b = 0$, $a \in (-\infty; -1) \cup \{0\} \cup (1; +\infty)$.
5. The case in Fig. 5.16 is possible when $a = -1/\sqrt{2}$, $b = 1/\sqrt{2}$ or $b = -1/\sqrt{2}$, $a = 1/\sqrt{2}$.

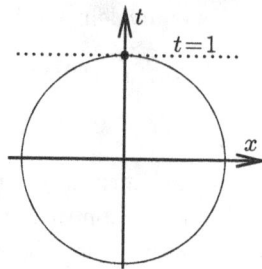

Fig. 5.12.   Case 1.

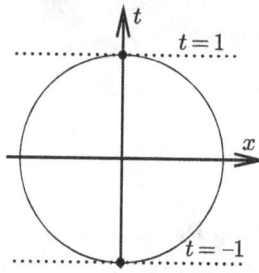

Fig. 5.13.   Case 2.

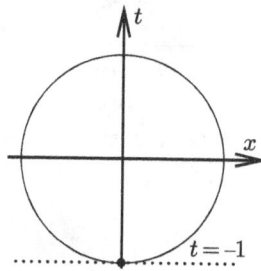

Fig. 5.14.   Case 3.

6. The case in Fig. 5.17 is possible when $a = 1/2$, $b = -1$ or $b = 1/2$, $a = -1$.

7. The case in Fig. 5.18 is possible when $a = -1/2$, $b = 1$ or $b = -1/2$, $a = 1$.

Answer: All possible values of a and b are described in Cases 1–7.   □

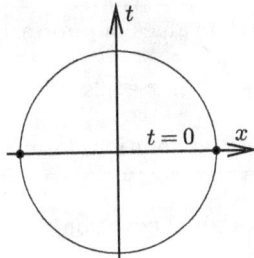

Fig. 5.15.   Case 4.

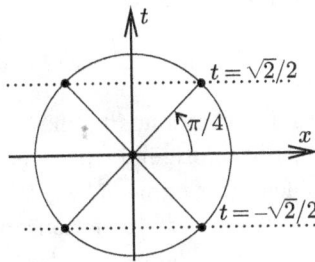

Fig. 5.16.   Case 5.

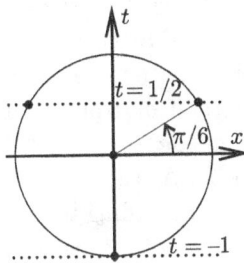

Fig. 5.17.   Case 6.

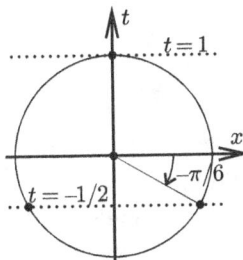

Fig. 5.18.   Case 7.

Example 5.25. Find all the values of a for each of which the equation

$$\cos 2x - 2a \sin x - |2a - 1| + 2 = 0$$

has solutions and all its positive solutions form an arithmetic progression.

Proof. We transform the original equation to the form

$$\sin^2 x + a \sin x + |a - 1/2| - 3/2 = 0.$$

Denoting $t = \sin x$, we obtain the quadratic equation

$$t^2 + a \cdot t + |a - 1/2| - 3/2 = 0. \tag{5.1}$$

Each root $\tilde{t}$ of this equation corresponds to, the equation $\sin x = \tilde{t}$, whose solutions are the solutions of the original equation.

In the previous problem (Example 5.24), it was determined when the set of solutions to a set of two equations of the form $\sin x = \tilde{t}$ forms an arithmetic progression. The sets of these solutions are shown in Figs. 5.12–5.18.

1. In the first case (see Fig. 5.12), the original equation is equivalent to the equation $\sin x = 1$. It is obtained from equation (5.1) if and only if one of the roots of (5.1) is the number 1, and the other root is either less than -1, or greater than or equal to 1. Substituting $t = 1$ into the equation, we get $1 + a + |a - 1/2| - 3/2 = 0$, from which we have $a \leq 1/2$, i.e., $a \in (-\infty; 1/2]$. At the same time, the equation takes the form $t^2 + at - a - 1 = 0$, and according to Vieta's theorem, its other root is $-a - 1$. Then, the above condition means that either

$$-a - 1 \geq 1 \Longrightarrow a \in (-\infty; -2]$$

   or

$$-a - 1 < -1 \Longrightarrow a \in (0; +\infty).$$

   Crossing the unions of these sets with the set $(-\infty; 1/2]$, we get part of the answer to the problem: $(-\infty; -2] \cup (0; 1/2]$.

2. In the second case (see Fig. 5.13), one of the roots of (5.1) is $t = 1$, and the other root must be either less than or equal to -1, or greater than 1.

When substituting $t = -1$ into equation (5.1), we get $1 - a + |a - 1/2| - 3/2 = 0$, from which we find $a = 0$. In this case, the equation takes the form $t^2 - 1 = 0$, and its second root $t = 1$ does not satisfy the conditions of the case under consideration. Therefore, the second case is impossible.

3. In the third case (see Fig. 5.14), the roots of equation (5.1) are the numbers $t = 1$ and $t = -1$, from which, according to Vieta's theorem, we get

$$\begin{cases} 1 + (-1) = -a, \\ 1 \cdot (-1) = |a - 1/2| - 3/2, \end{cases}$$

that is, $a = 0$. This value is part of the task response.

4. The fourth case (see Fig. 5.15) means that one of the roots of equation (5.1) is equal to 0, and the other is either also equal to 0, or less than -1, or more than 1. Substituting $t = 0$ into equation (5.1), we get $|a - 1/2| = 3/2$. This equation has two solutions, $a = -1$ and $a = 2$. If $a = -1$, then equation (5.1) takes the form $t^2 - t = 0$ and has the roots $t = 0$ and $t = 1$, and the second root does not satisfy the conditions of the case under consideration. If $a = 2$, then we get the equation $t^2 + 2t = 0$, whose second root $t = -2$ satisfies the conditions of this case. So, $a = 2$ is also part of the answer to the problem.

5. In the fifth case (see Fig. 5.16), equation (5.1) has the roots $t = -1/\sqrt{2}$ and $t = -1/\sqrt{2}$, and using Vieta's theorem, we get

$$\begin{cases} 1/\sqrt{2} + (-1/\sqrt{2}) = -a, \\ (1/\sqrt{2}) \cdot (-1/\sqrt{2}) = |a - 1/2| - 3/2. \end{cases} \iff \begin{cases} a = 0, \\ -1/2 = |a - 1/2| - 3/2. \end{cases}$$

This system is incompatible, and the case in question is impossible.

6. The sixth case (see Fig. 5.17) means that equation (5.1) has the roots $t = 1/2$ and $t = -1$, and again using Vieta's theorem, we get

$$\begin{cases} 1/2 + (-1) = -a, \\ (1/2) \cdot (-1) = |a - 1/2| - 3/2. \end{cases} \iff \begin{cases} a = 1/2, \\ -1/2 = |a - 1/2| - 3/2. \end{cases}$$

This system is also incompatible according to Vieta's theorem, and such a case is again impossible.

7. Finally, in the seventh case (see Fig. 5.18), $t = -1/2$, $t = 1$ and

$$\begin{cases} (-1/2) + 1 = -a, \\ (-1/2) \cdot 1 = |a - 1/2| - 3/2. \end{cases}$$

From here, we get $a = -1/2$, which gives another part of the answer.

It remains to combine the parts of the answer obtained in the first, third, fourth, and seventh cases.

Answer: $(-\infty; -2] \cup \{-1/2\} \cup [0; 1/2] \cup \{2\}$.   □

Problems

5.1. Solve the equation

$$\cos 2x + 6 \sin x - 5 = 0.$$

5.2. Solve the equation

$$\sqrt{3} \sin 2x + 2 \sin^2 x = 1.$$

5.3. Solve the equation

$$\sqrt{1 + \sin x} = 1 - 2 \sin x.$$

5.4. Solve the equation

$$\cos 4x = \cos^4 x - \sin^2 x.$$

5.5. Solve the equation

$$\cos\left(2x - \frac{\pi}{3}\right) - \sin x = \frac{1}{2}.$$

5.6. Solve the equation

$$\cos(\pi\sqrt{13 - x^2}) = 1/2.$$

5.7. Solve the equation $5 \sin x + 2 \cos 2x = 3.$

5.8. Solve the equation $\dfrac{4 \sin x - 3}{4 \sin^2 + \sin x - 3} = 2.$

5.9. Solve the equation

$$2 \sin^2 x = 4 \sin^2 2x + 7 \cos 2x - 6.$$

5.10. Solve the equation

$$\sqrt{11 - 8 \cos^4 x - 4 \sin x \cos x} = 3 \sin x + \cos x.$$

5.11. Solve the equation $\cos 3x \cdot \cos x = 1/8.$

5.12. Solve the equation

$$\sin 4x + 2\sin(5x/2)\cos(x/2) = 0.$$

5.13. Find the smallest positive root of the equation

$$8\cos(3x/4)\cos(x/4) - 6\cos x + 1 = 0.$$

5.14. Solve the equation

$$\sin 2x + \sin 3x + \cos 5x = 1.$$

5.15. Solve the equation

$$2\cos x - 3 = 6\sin x - 2\sin 2x.$$

5.16. Solve the equation

$$\cos 5x + \sin x \sin 4x = 0.$$

5.17. Solve the equation

$$\cos x \cos 3x - 9\cos^2 x + 5 = 14\sin x \sin 3x - 30\sin^2 x.$$

5.18. Solve the equation

$$\cos 0.3x - \cos 1.2x + \cos 0.9x = 1.$$

5.19. Solve the equation

$$\sin x (3\sin 2x \sin^3 x + 12\sin 2x \sin x - 16\cos x) + 2\sin 4x = 0.$$

5.20. Solve the equation

$$\cos x \sin\frac{x}{4} + \frac{9}{10}\sin x + 2\sin\frac{x}{4}\cos\frac{x}{2} + \sin\frac{x}{4}$$
$$-\frac{1}{2}\cos\frac{x}{4} - \frac{9}{20} = 0.$$

5.21. Solve the equation

$$|\cos 2x \sin 6x| + |\cos 6x \sin 2x| = \sin\frac{3\pi}{11}.$$

Tasks for Section 5.2

5.22. Solve the equation $\quad 5\cos x + 2\sin x = 3.$

5.23. Solve the inequalities

$$3\sin 2\pi x \geq \sqrt{2}\sin 4\pi x + 3\cos 2\pi x + \sqrt{32}.$$

5.24. Find the domain of definition values of the function

$$f(x) = \log_3(10\cos^2 3x - 14\sin 3x \cos 3x + 2 + \sqrt{3}).$$

5.25. Solve the equation

$$2\sqrt{2}\sin^2\left(x - \frac{\pi}{4}\right) - 5\sin x + 5\cos x + 2\sqrt{2} = 0.$$

5.26. Solve the equation

$$3(\sin x - 1) + 4\cos x + \cos\left(2x + 4\tan^{-1}\frac{1}{2}\right) = 0.$$

5.27. Solve the equation

$$\sqrt{3}\sin\left(3x - \frac{\pi}{5}\right) + 2\sin\left(8x - \frac{\pi}{3}\right)$$

$$= 2\sin\left(2x + \frac{11\pi}{15}\right) + 3\cos\left(3x - \frac{\pi}{5}\right).$$

5.28. Find all solutions of the equation

$$3\sin 5x + \sqrt{2}\cdot\sin^2\left(x + \frac{\pi}{20}\right) = 3\cos 5x - 3\sqrt{2}$$

on the segment $[-2\pi; 2\pi]$.

5.29. Solve the equation

$$\sin x + \sqrt{3}\cos x = 2 + 3\cos^2(2x + \pi/6).$$

5.30. Solve the inequality $\quad \sqrt{2\sin x} < 1.$

5.31. Solve the inequality $\quad 16\sin^2 x + \cot^2 x \leq 7.$

5.32. Solve the inequality $\quad \sin x \cdot \sin|x| \geq -1/2.$

5.33. Solve the inequality

$$\sqrt{4x - x^2 - 3}\cdot(\sqrt{2}\cos x - \sqrt{1 + \cos 2x}) \geq 0.$$

5.34. Find all the numbers x on the interval $[-\pi, \pi]$ satisfying the inequalities

$$3\cos x + 1 \le \cos 2x \le (4 - \sqrt{3})\sin x + 1 - 2\sqrt{3}.$$

Tasks for Section 5.3

5.35. Solve the equation

$$\sqrt{1 - \cos 2x} = \sqrt{2} \cdot \sin x \cdot (\cos x - 2/3).$$

5.36. Solve the equation

$$\tan 8x - \tan 6x = \frac{1}{\sin 4x},$$

for $x \in [-\pi/4; 3\pi/4]$.

5.37. Solve the equation

$$\log_2\left(\cos 3\left(\frac{\pi}{6} - x\right)\right) \cdot \log_2(\cos 2x) + \log_2\left(\sin 5x + \sin x\right) = 0.$$

5.38. Find all solutions of the equation

$$2\cos\frac{x}{3} + 2(\sqrt{5} - 1)\sin\frac{x}{6} = 2 - \sqrt{5}$$

satisfying the condition $\cos(3x/4) < 0$.

5.39. Solve the equation

$$\tan x + \tan 3x + \tan x \tan 3x \tan 4x = \tan 4x + \tan 5x.$$

5.40. Solve the equation

$$16\cos^2 x \cdot \left(\cot^2(2x) - 1\right) \cdot \cos 4x = \frac{1}{\sin^2 4x}.$$

5.41. Solve the equation

$$4\sin 2x - 3\cos 2x - 4\sin x - 2\cos x + 1 = 0.$$

5.42. Solve the equation

$$3\sin^2 x - 3\cos x - 6\sin x + 2\sin 2x + 3 = 0.$$

5.43. Solve the equation

$$\sqrt{\sin 2x} = \sqrt{\cos x - \sin x - 1}.$$

5.44. Solve the equation

$$\sqrt{-3\sin 2x} = -2\sin 2x - \sin x + \cos x - 1.$$

5.45. Solve the equation

$$\cos 6x - 4\cos 5x + \cos 4x - 5\cos x + 7 = 0.$$

5.46. Solve the equation

$$\frac{4}{3}\cos^4 x + \frac{1}{2}\sin^2 x = \frac{1}{4}\cos 2x + \cos^2 x \cos 12x.$$

5.47. Solve the equation

$$\cos x(\cos 2x + \cos 4x) + \sin^2 x = 2.$$

Tasks for Section 5.4

5.48. Find all the values of a such that the function

$$y(x) = \log_{25-a^2}(\cos x + \sqrt{8}\sin x - a)$$

is defined for all values of x.

5.49. For what values of a the inequality

$$\log_{(2a-15)/5}\left(\frac{\sin x + \sqrt{3}\cos x}{5} + \frac{a}{5} - 1\right) > 0$$

is executed for all x?

5.50. For what values of a does the equation

$$2\cos^2\left(2^{2x-x^2}\right) = a + \sqrt{3}\sin\left(2^{2x-x^2+1}\right)$$

have at least one solution?

5.51. Find all the values of a for which, among the roots of the equation

$$\sin 2x + 6a\cos x - \sin x - 3a = 0,$$

there are two roots, the difference between which is $3\pi/2$.

5.52. Find all the values of the parameter a for each of which the equation

$$(a^2 - 6a + 9)(2 + 2\sin x - \cos^2 x)$$
$$+ (12a - 18 - 2a^2)\cdot(1 + \sin x) + a + 3 = 0$$

has no solutions.

5.53. Find all the values of the parameter a for which the inequality

$$|3\sin^2 x + 2a\sin x \cos x + \cos^2 x + a| \le 3$$

is executed for any value of x.

5.54. For each value of b, solve the equation

$$3\cos x \sin b - \sin x \cos b - 4\cos b = 3\sqrt{3}.$$

5.55. Find all the valid values of the parameter a for which the range of values of the function $y = \frac{\sin x + 2(1-a)}{a - \cos^2 x}$ contains the segment $[1; 2]$.

5.56. Solve the equation

$$\frac{3\cos x + 2\sin x}{\cos x} = \frac{\cos 2x}{\cos^2 x} + \frac{\cos x + \sin x}{\cos x}$$

$$\cdot \sqrt{3 + 2x - 2y + 2xy - x^2 - y^2}.$$

5.57. Find all the values of the parameter a for which any root of the equation

$$a(2a - 1)\sin^3 x + 3\cos^3 x - 2a^2 \sin x = 0$$

is the root of the equation

$$\log_{1/2}(3\tan x - 1) - \log_2(3\tan x + 1) - \log_{1/\sqrt{2}}(5 - \tan x) = 1$$

and, conversely, any root of the second equation is the root of the first equation.

5.58. Find all the values of the parameter a such that the system of equations

$$\begin{cases} \cos x = \sin(x\sqrt{4 - 7a^2}), \\ \sin x = \left(3a - \dfrac{1}{2}\right)\cos(x\sqrt{4 - 7a^2}) \end{cases}$$

has exactly one solution on the segment $[\pi/2; 5\pi/2]$.

5.59. Let t_1 and t_2 be the roots of the quadratic equation

$$t^2 - (5b - 2)^2 t - 3b^2 - 7b + 1 = 0.$$

Find all the values of the parameter b such that for any value of the parame- ter a, the function

$$f(x) = \cos(a\pi x) \cdot \cos((t_1^3 + t_2^3) \cdot \pi x)$$

is periodic.

5.60. Find all the values of a for which the inequality

$$\log_5 (a \cos 2x - (1 + a^2 - \cos^2 x) \cdot \sin x + 4 - a) \le 1$$

is executed for all x.

5.61. Find all the values of a for which there are exactly six roots of the following equation on the segment $[\pi/2; 3\pi/2]$:

$$\cos 6x + a = (2a + 1) \cos 3x.$$

5.62. Find all the values of a for which the equation

$$(|a| - 1) \cos 2x + (1 - |a - 2|) \sin 2x$$
$$+ (1 - |2 - a|) \cos x + (1 - |a|) \sin x = 0$$

has an odd number of different solutions on the interval $(-\pi; \pi)$.

5.63. Find all the values of a for which the distance between any neighboring roots of the equation

$$3 \tan a \cdot \cos 2x + 3\sqrt{2} \cos 3a \cdot \cos x + 3 \tan a - \cot a = 0,$$

is less than or equal to $\pi/2$.

5.64. Find all the values of a for which, for any root of the equation

$$\cos a \cos 3x - \sin 3a \cos x + 2 \sin 2a \cos 2x = 3 \sin a - \cos 3x,$$

there is another root at a distance of at most $\pi/3$ from it.

5.65. Find all the values of a such that the equation

$$\cos 2x + 2a \cos x + |2a + 1| - 2 = 0$$

has solutions and all its positive solutions form an arithmetic progression.

Hints and Answers

5.1. $x = \pi/2 + 2\pi n$, $n \in \mathbb{Z}$.

5.2. $\pi/12 + \pi n/2$, $n \in \mathbb{Z}$.

5.3. πn, $n \in \mathbb{Z}$.

5.4. $\pi n/3$, $n \in \mathbb{Z}$.

5.5. πm, $\pm\pi/3 - \pi/6 + 2\pi n$, $m, n \in \mathbb{Z}$.

5.6. $92/3$, $\pm\sqrt{13 - (1/3 + 2n)^2}$, $n = 0, 1$.

5.7. $x = \pi/2 + 2\pi m$, $x = (-1)^n \arcsin(1/4) + \pi n$, $m, n \in \mathbb{Z}$.

5.8. $x = (-1)^{n+1}\pi/6 + \pi n$, $n \in \mathbb{Z}$.

5.9. $x = \pm\pi/6 + \pi n$, $n \in \mathbb{Z}$.

5.10. $x = \pi/12 + 2\pi m$, $x = 5\pi/12 + 2\pi n$, $m, n \in \mathbb{Z}$.

5.11. $\left(\pm \arccos\left(\frac{-1+\sqrt{11}}{4}\right) + 2\pi n\right)/2$, $n \in \mathbb{Z}$.

5.12. $\pi n/3$, $n \in \mathbb{Z}$.

5.13. $x = 4\pi/3$.

5.14. $x = 2l\pi/5$, $\pi/4 + m\pi$, $\pi/6 + 2n\pi/3$, $l, m, n \in \mathbb{Z}$.

5.15. $(-1)^{n+1}\pi/6 + \pi n$, $n \in \mathbb{Z}$.

5.16. $\pi/2 + \pi n$, $\pi/8 + \pi m/4$, $n, m \in \mathbb{Z}$.

5.17. $\pm(1/2) \cdot \arccos(2/5) + \pi n$, $n, m \in \mathbb{Z}$.

5.18. $5\pi/6 + 5\pi m/3$, $20\pi n/9$, $n, m \in \mathbb{Z}$.

5.19. $\pi m/2$, $\pm\frac{1}{2}\arccos\left(\frac{-1}{3}\right) + \pi n$, $m, n \in \mathbb{Z}$.

5.20. $-23\pi/6$, $-19\pi/6$, $-11\pi/6$, $-4\arccos(-9/10)$. **Hint:** Group and simplify the expression when $\sin(x/4)$.

5.21. $\pm 3\pi/88 + \pi n/4$, $n \in \mathbb{Z}$. **Hint:** Use the fact that the solution of the equation $|a| + |b| = c$ is equivalent to two equations: $|a + b| = c$ if $ab \geq 0$ and $|a - b| = c$ if $ab \leq 0$.

5.22. $\arccos(5/\sqrt{29}) \pm \arccos(3/\sqrt{29}) + 2\pi n$, $n \in \mathbb{Z}$.

5.23. $3/8 + n$, $n \in \mathbb{Z}$.

5.24. $\mathbb{R}$.

5.25. $\pi/4 + (-1)^n\pi/6 + \pi n$, $n \in \mathbb{Z}$.

5.26. $-\tan^{-1}(4/3) + (-1)^n(\pi/6) + \pi n$, $n \in \mathbb{Z}$.

5.27. $8\pi/45 + \pi m/3$, $-\pi/25 \pm \pi/6 + 2\pi n/5$, $m, n \in \mathbb{Z}$.

5.28. $-\pi/20$, $39\pi/20$.

5.29. $\pi/6 + 2\pi n$, $n \in \mathbb{Z}$.

5.30. $2\pi n \le x < \pi/6 + 2\pi n$; $5\pi/6 + 2\pi n < x \le \pi + 2\pi n$, $n \in \mathbb{Z}$.

5.31. $\pm\pi/6 + \pi n$, $n \in \mathbb{Z}$.

5.32. $x \in [-\pi/4 - \pi n; \pi/4 - \pi n] \cup (-1/4; +\infty)$, $n \in \mathbb{N}$.

5.33. $x \in [1; \pi/2] \cup \{3\}$.

5.34. $2\pi/3$.

5.35. $\pi m, -\arccos(-1/3) + 2\pi n$; $m, n \in \mathbb{Z}$.

5.36. $\pi/8 + \pi n/4$, $n = -1, 0, 1, 2$; $-\pi/4 + \pi k/10$, $k = 1, 2, 3, 4, 6, 7, 8, 9$.

5.37. $\pi/6 + 2\pi k$, $5\pi/6 + 2\pi l$, $\pi/18 + 2\pi m$, $\pi/18 + 2\pi n$, $k, l, m, n \in \mathbb{Z}$.

5.38. $-\pi + 24\pi m$, $7\pi + 24\pi n$, $m, n \in \mathbb{Z}$.

5.39. $\pi n/5$, $n \in \mathbb{Z}$.

5.40. $\pi n/7$, $\pi m/9$, $n, m \in \mathbb{Z}$, $n \ne 7l$, $m \ne 9l$, $l \in \mathbb{Z}$.

5.41. Hint: Use a replacement: $t = 2\sin x + \cos x$.

5.42. Hint: Use a replacement: $t = 2\sin x + \cos x$. $2\pi n$, $\pi/2 + 2\pi k$, $2\arccos(1/\sqrt{5}) + 2\pi l$, $2\arccos(1/\sqrt{5}) - \pi/2 + 2\pi m$, $n, k, m, l \in \mathbb{Z}$.

5.43. Hint: Use a replacement: $t = \cos x - \sin x$. $-\pi/4 \pm \pi/4 + 2\pi n$, $n \in \mathbb{Z}$.

5.44. Hint: Use a replacement: $t = \sin x - \cos x$. $\pi/4 + (-1)^{m+1}\pi/4 + \pi m$, $\pi/4 + (-1)^{n+1}\pi/3 + \pi n$, $m, n \in \mathbb{Z}$.

5.45. $2\pi n$, $n \in \mathbb{N}$.

5.46. $x = \pm\pi/6 + \pi n$, $n \in \mathbb{Z}$.

5.47. $2\pi n$, $n \in \mathbb{Z}$.

5.48. $a \in (-5; -\sqrt{24}) \cup (-\sqrt{24}; -3)$.

5.49. $a \in (15/2; 8) \cup (12; +\infty)$.

5.50. $a \in [-1; 2)$.

5.51. $a \in \{\pm 1/6, \pm\sqrt{2}/6\}$.

5.52. $a \in (-\infty; -3) \cup (1; 6)$.

5.53. $a \in [-12/5; 0]$.

5.54. If $b = 5\pi/6 + 2\pi l$, $l \in \mathbb{Z}$, then $x = \pi/6 + 2\pi k$, $k \in \mathbb{Z}$; if $b = -5\pi/6 + 2\pi m$, $m \in \mathbb{Z}$, then $x = 5\pi/6 + 2\pi n$, $n \in \mathbb{Z}$; with other parameter values b, there are no solutions.

5.55. $a \in [1/3; 3/4) \cup (3/4; 33/32]$. **Hint:** Rewrite the condition in the form of inequalities.

5.56. $x = \pi n$, $y = \pi n - 1$, $n \in \mathbb{Z}$. **Hint:** Consider the variable $t = \tan x$ and examine the root expression.

5.57. $a = 1$. **Hint:** Solve the second equation.

5.58. $-1/6$, $-1/2$, $\pm 3/4$, $\pm 1/4\sqrt{39/7}$.

5.59. $b = 2/5$.

5.60. $a \in [0; 1)$.

5.61. $a \in (-2/3; 0)$.

5.62. $a \in [0; 1) \cup (1; 2] \cup \{3\}$. **Hint:** Imagine the equation as $A\cos 2x + B\sin 2x = -B\cos x + A\sin x$, and solve it by introducing an auxiliary argument (one angle φ for expressions in different parts of the equation).

5.63. $a = \pm\pi/6 + \pi n$, $n \in \mathbb{Z}$.

5.64. $a = \pi n$, $n \in \mathbb{Z}$.

5.65. $a \in \{-2\} \cup [-1/2; 0] \cup \{1/2\} \cup [2; +\infty)$.

Chapter 6

Tasks for the Number of Solutions

6.1 Problems for the Uniqueness of the Solution or for an Odd Number of Solutions for Even Functions

The notation $f(a, x)$ means that the function depends on the parameter a and the variable x. The main types of tasks in this section can be formulated as follows:

Task A. *Find all the values of the parameter a (or several parameters) for which the equation (or inequality) $f(a, x) = 0$ ($f(a, x) \leq 0$ or $f(a, x) \geq 0$) has a single solution.*

Recall the definitions of even and odd functions.

Definition 6.1. If the domain of the definition of the function $f(x)$ is symmetric with respect to the origin and if for each x from the domain of definition the equality is satisfied $f(-x) = f(x)$, then the function $f(x)$ is *even*. If the domain of definition is symmetric with respect to the origin and for each x from the domain of definition the equality $f(-x) = -f(x)$ is fulfilled, then the function $f(x)$ is *odd*.

So, for example, the functions

$$f_1(x) = |\sin x|,$$

$$f_2(x) = \cos x,$$

$$f_3(x) = x^4 - 3x^2,$$

$$f_4(x) = \frac{\tan x \cdot (7^x - 1)}{7^x + 1}$$

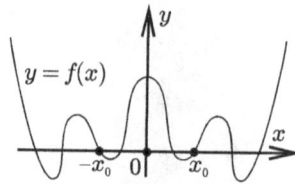

Fig. 6.1. $f(\pm x_0) = 0$.

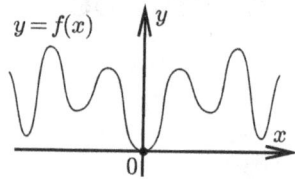

Fig. 6.2. $f(\pm x_0) = f(0) = 0$.

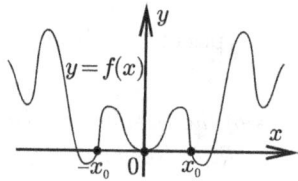

Fig. 6.3. $f(0) = 0$.

are even. For the first three, this is obvious. Let us check that the function $f_4(x)$ is even:

$$f_4(-x) = \frac{\tan(-x) \cdot (7^{-x} - 1)}{7^{-x} + 1} = \frac{-\tan x \cdot 7^{-x}(1 - 7^x)}{7^{-x}(1 + 7^x)}$$

$$= -\frac{\tan x \cdot (1 - 7^x)}{(1 + 7^x)} = \frac{\tan x \cdot (7^x - 1)}{7^x + 1} = f_4(x).$$

Let us assume that when solving problem A, the function $f(a, x)$ turns out to be even for each value of a. Then, if x_0 is the solution to problem A, then $-x_0$ is the solution to problem A (see Fig. 6.1) since $f(a, x_0) = f(a, -x_0)$. Therefore, for the uniqueness of the solution, it is *necessary* for $x_0 = 0$ to be the solution to problem A (see Figs. 6.2 and 6.3), and it is *enough* that there are no more solutions (except $x_0 = 0$) (see Fig. 6.3), so we discard the case depicted in Fig. 6.2.

In solving problem A, we:

1. search for possible values of a from the equation (inequality) $f(a,0) = 0$ ($f(a,0) \leq 0$, $f(a,0) \geq 0$), i.e., from the necessary condition of uniqueness of the solution tasks;
2. for the values of a found from the necessary condition, check that there are no other solutions (except $x = 0$), i.e., the sufficient condition of uniqueness of the solution.

Example 6.1. Find all values of a for which the inequality

$$\cos x - 2\sqrt{x^2 + 9} \leq -\frac{x^2 + 9}{a + \cos x} - a$$

has a unique solution.

Proof. Transform the inequality

$$\cos x - 2\sqrt{x^2 + 9} \leq -\frac{x^2 + 9}{a + \cos x} - a$$

$$\Longleftrightarrow \frac{(a + \cos x)^2 - 2\sqrt{x^2 + 9}(a + \cos x) + x^2 + 9}{a + \cos x} \leq 0$$

$$\Longleftrightarrow \frac{(a + \cos x - \sqrt{x^2 + 9})^2}{a + \cos x} \leq 0.$$

Denote

$$f(x) = \frac{(a + \cos x - \sqrt{x^2 + 9})^2}{a + \cos x}.$$

Since $f(x)$ is an even function ($f(x) = f(-x)$), then in order for the original inequality $f(x) \leq 0$ to have a unique solution, it is necessary that $x = 0$ is the solution of the inequality (since if x_0 is the solution of the inequality, then $-x_0$ is its solution due to the fact that the function $f(x)$ is even).

Thus, $(a-2)^2/(a+1) \leq 0$, i.e., $a = 2$ and $a < -1$. Let us check whether the solution $x = 0$ of the original inequality is the only one for the values of a found by us.

1. Let $a < -1$ (see Fig. 6.4 for $a = -2$). Then, the inequality

$$\frac{(a + \cos x - \sqrt{x^2 + 9})^2}{a + \cos x} \leq 0$$

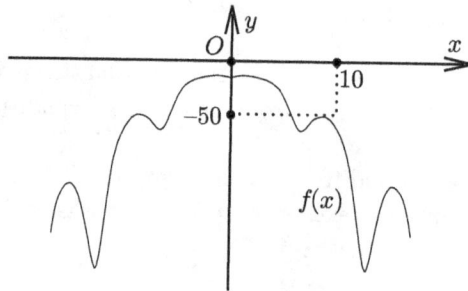

Fig. 6.4.   Function $f(x)$ for $a = -2$.

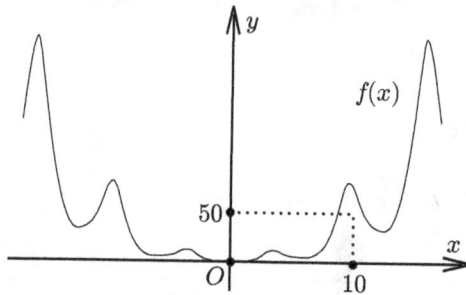

Fig. 6.5.   Function $f(x)$ for $a = 2$.

is fulfilled for all $x \in \mathbb{R}$ since for $a < -1$, the following inequality holds:

$$(a + \cos x - \sqrt{x^2 + 9})^2 \geq 0, \quad a + \cos x < 0.$$

2. Let $a = 2$ (see Fig. 6.5). Then, $2 + \cos x > 0$, for any $x \in \mathbb{R}$. Therefore,

$$\frac{(2 + \cos x - \sqrt{x^2 + 9})^2}{2 + \cos x} \leq 0 \Longleftrightarrow (2 + \cos x - \sqrt{x^2 + 9})^2 \leq 0$$

$$\Longleftrightarrow 2 + \cos x - \sqrt{x^2 + 9} = 0 \Longleftrightarrow 2 + \cos x = \sqrt{x^2 + 9}.$$

But $x = 0$ is the only root of the equation $2 + \cos x = \sqrt{x^2 + 9}$ because, for $x \neq 0$, we get an incorrect inequality:

$$3 < \sqrt{x^2 + 9} = 2 + \cos x \leq 3.$$

Answer: $a = 2$.

□

Example 6.2. Find all the values of a for which the following system has a unique solution:

$$\begin{cases} (x^2 + 1)a = y - \cos x, \\ \sin^4 x + |y| = 1. \end{cases}$$

Proof. The system is even with respect to x since each equation in the system contains even functions of the variable x. Therefore, with each solution $(x_0; y_0)$ of our system, there will be a solution $(-x_0; y_0)$. Therefore, for the uniqueness of the solution, the equality $x_0 = 0$ is necessary. Then, $a = y - 1$ and $y = \pm 1$, from which we get $a = 0$ or $a = -2$. It is easily checked that for $a = 0$, there are infinitely many solutions. When $a = -2$, the system takes the form

$$\begin{cases} y = \cos x - 2(x^2 + 1), \\ \sin^4 x + 2(x^2 + 1) - \cos x = 1. \end{cases}$$

However, from the equation $\sin^4 x + 2x^2 + (1 - \cos x) = 0$, we conclude that the solution $x = 0$ is unique. Therefore, the system has a single solution: $(x, y) = (0; -1)$.

Answer: $a = 2$. $\qquad\qquad\qquad\qquad\qquad\qquad\qquad\qquad\qquad\qquad$ □

Example 6.3. For what values of a and b does the system

$$\begin{cases} \left| \dfrac{x^y - 1}{x^y + 1} \right| = a, \\ x^2 + y^2 = b \end{cases}$$

have a single solution?

Proof. First, we find all possible parameter values at which a single solution of the system is possible, and then we check for the number of solutions and select the parameter values we need.

1. Let $f(x, y) = \left| \dfrac{x^y - 1}{x^y + 1} \right|$, $g(x, y) = x^2 + y^2$. From the equalities $f(x, y) = f(x, -y)$, $g(x, y) = g(x, -y)$, it follows that if $(x_0; y_0)$ is the solution to the original system, then $(x_0; -y_0)$ is also the solution to the system.

Therefore, for the uniqueness of the solution, the condition $y_0 = -y_0$ must be fulfilled, i.e., $y_0 = 0$. Substituting $y_0 = 0$ into the original system, we get the system

$$\begin{cases} a = 0, \\ x^2 = b. \end{cases}$$

2. So, the number a is zero. Let us find out for which b the system

$$\begin{cases} x^y = 1, \\ x^2 + y^2 = b \end{cases}$$

obtained from the original one at $a = 0$ has a unique solution. This system is defined at $x > 0$, and at the same time, it is equivalent to the set of systems

$$\left[\begin{array}{l} \begin{cases} y = 0, \\ x = \sqrt{b} \quad \text{(if } b > 0\text{)}, \end{cases} \\ \begin{cases} x = 1, \\ y = \pm\sqrt{b-1} \quad \text{(if } b \geq 1\text{)}. \end{cases} \end{array} \right.$$

Solving this set, we find that:

- there are no solutions for $b \leq 0$;
- for $b \in (0; 1]$, there is one solution: $(x; y) = (\sqrt{b}; 0)$;
- for $b > 1$, there are three solutions: $(x; y) = (\sqrt{b}; 0), (1; \pm\sqrt{b-1})$.

Answer: $a = 0$, $b \in (0; 1]$. ☐

6.2   Problems on the Number of Solutions Involving Symmetries

This section is a continuation of the previous one.

1. In the previous section, symmetry with respect to the line $x = 0$ (the concept of an even function) was considered. Now, we consider symmetries in a more general situation, in particular, we consider symmetries with respect to the lines $x = b$, where b is some given number.

   In problems of this kind, it is convenient to make the substitution $z = x - b$. If there is symmetry with respect to the line $x = b$, where b

is some given number, the function $f(z) = f(x - b)$ will be even relative to the new variable: $f(-z) = f(z)$.

2. When solving, for example, an equation of the form $f(x,y) = 0$, it may turn out that for all possible values of x, y, the equality $f(x,y) = f(y,x)$ holds (symmetry with respect to the bisector of the first coordinate angle). Then, together with the solution $(x_0; y_0)$ of this equation, the pair $(y_0; x_0)$ will also be its solution. For the uniqueness of the solution in this case, it is necessary to fulfil the equality $x = y$.

Example 6.4. Find all the values of a for each of which the system of inequalities

$$\begin{cases} y \geq x^2 + 2a, \\ x \geq y^2 + 2a \end{cases}$$

has a single solution.

Proof. Let $(x_0; y_0)$ be the solution of the system. Then, due to symmetry, $(y_0; x_0)$ will also be a solution. Therefore, the equality of $x = y$ is a necessary condition for the uniqueness of the solution. Substituting it into the system, we get

$$x^2 - x + 2a \leq 0.$$

If this inequality has two or more solutions, then the original system has at least two solutions, and this case does not suit us. If the inequality has no solutions, then the original system has an even number of solutions, an infinite number of solutions, or no solutions, but all these cases do not suit us. Let this inequality have a unique solution, then the discriminant of the quadratic equation vanishes, i.e.,

$$D = 1 - 8a = 0 \Longleftrightarrow a = 1/8,$$

and $x = y = 1/2$. Let us check the sufficiency of this condition. Adding the two inequalities, we get

$$x + y \geq x^2 + y^2 + 1/2 \Longleftrightarrow (x - 1/2)^2 + (y - 1/2)^2 \leq 0.$$

Therefore, the solution $x = y = 1/2$ is really the only one.

Answer: For $a = 1/8$, the system of inequalities has a unique solution $x = y = 1/2$. $\qquad\square$

Example 6.5. Find all the values of the parameter a for which the equation

$$2^{-x^2} \cdot 4^x + \sin(\pi x/4) + \cos(\pi x/4) - 2 = a^3 - 3a^2 + a + \sqrt{2}$$

has a unique solution.

Proof. We transform the equation using the relations

$$\sin\left(\frac{\pi x}{4}\right) + \cos\left(\frac{\pi x}{4}\right) = \sqrt{2}\left(\cos\left(\frac{\pi x}{4}\right) \cdot \cos\left(\frac{\pi}{4}\right) + \sin\left(\frac{\pi x}{4}\right) \cdot \sin\left(\frac{\pi}{4}\right)\right)$$

$$= \sqrt{2}\left(\cos\left(\frac{\pi x}{4} - \frac{\pi}{4}\right)\right)$$

and $2^{-x^2} \cdot 4^x = 2^{-x^2+2x} = 2^{-(x-1)^2+1}$. We get

$$2 \cdot 2^{-(x-1)^2} + \sqrt{2}\cos\left(\frac{\pi(x-1)}{4}\right) - 2 - \sqrt{2} = a^3 - 3a^2 + a.$$

Let's replace $t = x - 1$ and denote $f(t) = 2 \cdot 2^{-t^2} + \sqrt{2}\cos(\pi t/4) - 2 - \sqrt{2}$. Then, the original problem is equivalent to finding all the values of the parameter a for which the equation $f(t) = a^3 - 3a^2 + a$ has a unique solution. However, since $f(t) = f(-t)$, i.e., the function $f(t)$ is even, and $f(t) < f(0)$, $t \neq 0$, the problem has a unique solution if and only if $t = 0$ is the solution of the equation $f(t) = a^3 - 3a^2 + a$. Substituting $t = 0$ into this equation and noting that $f(0) = 2 + \sqrt{2} - 2 - \sqrt{2} = 0$, we get that it is enough to solve the equation $a^3 - 3a^2 + a = 0$.

Answer: $a = 0$, $a = (3 \pm \sqrt{5})/2$. $\qquad\qquad\qquad\qquad\qquad\qquad\square$

Example 6.6. Find all rational values of a for which the equation

$$\frac{2(1-2a)}{\pi}\left(\arcsin\frac{2x}{x^2+1}\right) + a^2\left(\tan^{-1}x - \tan^{-1}\frac{1}{x}\right)^2 + a^2 + 3a - 3 = 0$$

has a single solution.

Proof. Let us introduce the notation

$$f(x) = \frac{2(1-2a)}{\pi}\left(\arcsin\frac{2x}{x^2+1}\right) + a^2\left(\tan^{-1}x - \tan^{-1}\frac{1}{x}\right)^2.$$

1. For the function $f(x)$, the equality $f(x) = f(1/x)$ holds, so if x_0 is the solution of the equation, then $1/x_0$ is also a solution. Therefore, an odd number of solutions (the only solution in our case) is possible only under the condition

$$x_0 = \frac{1}{x_0} \iff x_0^2 = 1 \iff x_0 = \pm 1.$$

Substituting $x = 1$ into the original equation, we get

$$a^2 + a - 2 = 0 \iff a_1 = 1; \ a_2 = -2.$$

Substituting $x = -1$ into the original equation, we get

$$a^2 + 5a - 4 = 0 \iff a_3 = (-5 + \sqrt{41})/2; \ a_4 = (-5 - \sqrt{41})/2.$$

The values of $a_{3,4}$ are irrational, so they do not satisfy the condition of the problem. Consider $a_{1,2}$.

2. Let us find out for which of the found values of a the equation has a unique solution.

- Let $a = 1$, then the equation takes the form

$$-\frac{2}{\pi}\left(\arcsin \frac{2x}{x^2+1}\right) + \left(\tan^{-1} x - \tan^{-1} \frac{1}{x}\right)^2 + 1 = 0$$

$$\iff \left(\tan^{-1} x - \tan^{-1} \frac{1}{x}\right)^2 + 1 = \frac{2}{\pi}\left(\arcsin \frac{2x}{x^2+1}\right).$$

Since $\dfrac{2|x|}{x^2+1} \le 1$,

$$1 \le \left(\tan^{-1} x - \tan^{-1} \frac{1}{x}\right)^2 + 1 = \frac{2}{\pi}\left(\arcsin \frac{2x}{x^2+1}\right) \le 1.$$

Therefore, in order for the equation to hold, it is necessary and sufficient that the following conditions hold:

$$\begin{cases} \tan^{-1} x - \tan^{-1} \dfrac{1}{x} = 0, \\ \dfrac{2}{\pi}\left(\arcsin \dfrac{2x}{x^2+1}\right) = 1. \end{cases} \iff x = 1.$$

Thus, for $a = 1$, the solution of the original equation is unique.

- Let $a = -2$, then the function $f(x)$ takes the form

$$f(x) = \frac{10}{\pi} \left(\arcsin \frac{2x}{x^2 + 1} \right) + 4 \cdot \left(\tan^{-1} x - \tan^{-1} \frac{1}{x} \right)^2,$$

and the original equation takes the form $f(x) = 5$. The inequality is true: $f(-1) = -5 < 5$. Further, if x tends to zero, with the remaining less than zero, then $\tan^{-1} x$ tends to 0, $\tan^{-1}(1/x)$ to $-\pi/2$, $\arcsin \frac{2x}{x^2+1}$ tends to 0. Therefore, $f(x)$ tends to π^2, and $\pi^2 > 5$. We have shown that the function $f(x)$ is continuous on the interval $(-2; 0)$ and takes values both greater and less than 5. There is a number $x_0 \in (-2; 0)$ such that $f(x_0) = 5$; hence, it follows that the original equation for $a = -2$ has at least two solutions: $x = 1$ and $x = x_0$. (With a more detailed consideration of this equation, it can be shown that it will have exactly five solutions for $a = -2$.)

Answer: For $a = 1$, the system has a unique solution of $x = 1$. □

Example 6.7. For what values of a the equation

$$|x| + \left| \frac{x+1}{3x-1} \right| = a$$

has exactly three different solutions?

Proof. Let us introduce the notation

$$f(x) = |x| + \left| \frac{x+1}{3x-1} \right|.$$

1. The following equality is true:

$$\frac{\left(\frac{x+1}{3x-1} \right) + 1}{3 \cdot \left(\frac{x+1}{3x-1} \right) - 1} = \frac{(x+1) + (3x-1)}{3(x+1) - (3x-1)} = \frac{4x}{4} = x.$$

It follows from it that if x_0 is the solution of the equation, then and $x_1 = \frac{x_0+1}{3x_0-1}$ is also the root of the original equation because $f(x_0) = f(x_1)$, which implies that an odd number of solutions is possible only if

$$x_0 = \frac{x_0 + 1}{3x_0 - 1} \iff 3x_0^2 - 2x_0 - 1 = 0 \iff$$

$$\iff x_0 = 1, \ x_0 = -1/3,$$

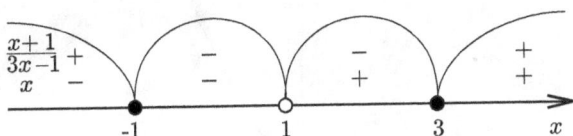

Fig. 6.6.   Method of intervals.

that is, when the roots of x_0 and x_1 coincide. Find the values of a that correspond to the values of $x_0 = 1$ and $x_0 = -1/3$:

$$a_1 = f(1) = 2, \quad a_2 = f(-1/3) = 2/3.$$

2. Let us check whether the equation will have exactly three solutions for the values of a found. Let $a = 2$. Let's solve the equation $f(x) = 2$. To do this, consider the four intervals $(-\infty; -1) \cup [-1; 0] \cup (0; 1/3) \cup (1/3; +\infty)$, and solve the equation $f(x) = 2$ on each of them (see Fig. 6.6).

• Let $x \in (1/3; +\infty)$. Then, the equation $f(x) = 2$ takes the form

$$x + \frac{x+1}{3x-1} = 2 \Longleftrightarrow 3x^2 - x + x + 1 = 2(3x-1)$$

$$\Longleftrightarrow x^2 - 2x + 1 = 0 \Longleftrightarrow x = 1.$$

• Let $x \in (0; 1/3)$. Then, the equation $f(x) = 2$ takes the form

$$x - \frac{x+1}{3x-1} = 2 \Longleftrightarrow 3x^2 - x - x - 1 = 2(3x-1)$$

$$\Longleftrightarrow 3x^2 - 8x + 1 = 0 \Longleftrightarrow x = \frac{4 \pm \sqrt{13}}{3}.$$

Only one root, $x = (4-\sqrt{13})/3$, belongs to the interval $(0; 1/3)$. Thus, we have found the second root of the equation as $f(x) = 2$.
• Let $x \in [-1; 0]$. Then, the equation $f(x) = 2$ takes the form

$$-x - \frac{x+1}{3x-1} = 2 \Longleftrightarrow 3x^2 - x + x + 1 = -2(3x-1)$$

$$\Longleftrightarrow 3x^2 + 6x - 1 = 0 \Longleftrightarrow x = -1 \pm \frac{2\sqrt{3}}{3}.$$

However, since $2\sqrt{3}/3 > 1$, each of the numbers $-1 \pm 2\sqrt{3}/3$ does not belong to the segment $[-1; 0]$.

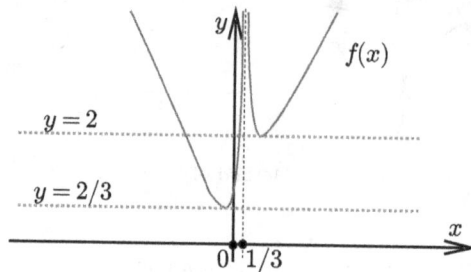

Fig. 6.7.   Graph of $f(x)$.

- Let $x \in (-\infty; -1)$. Then, the equation $f(x) = 2$ takes the form

$$- x + \frac{x+1}{3x-1} = 2 \Longleftrightarrow -3x^2 + x + x + 1 = 2(3x-1)$$

$$\Longleftrightarrow 3x^2 + 4x - 3 = 0 \Longleftrightarrow x = \frac{-2 \pm \sqrt{13}}{3}.$$

Only one root, $x = (-2 - \sqrt{13})/3$, belongs to the semi-axis $(-\infty; -1)$. Thus, we have found the third root of the equation $f(x) = 2$.

So, for $a = 2$, we checked that there are really exactly three solutions. Similarly, it can be proved that in the case of $a = 2/3$, the equation $f(x) = 2$ will have one solution (see Fig. 6.7). Therefore, only one value will be included in the task response: $a = 2$.

Answer: $a = 2$. $\square$

Example 6.8. Find all the values of a for which the system

$$\begin{cases} z\cos(x-y) + (2+xy)\sin(x+y) - z = 0, \\ x^2 + (y-1)^2 + z^2 = a + 2x, \\ (x+y+a\sin^2 z)\,((1-a)\ln(1-xy)+1) = 0 \end{cases}$$

has a single solution.

Proof.

1. Note that if $(x; y; z)$ is the solution of the system, then $(y; x; z)$ is also the solution of this system. For the uniqueness of the solution, the equality

$x = y$ is necessary. In this case, the system takes the form

$$
\begin{cases}
(2 + x^2)\sin 2x = 0, \\
2(x - 1)^2 + z^2 = a + 1, \\
(2x + a\sin^2 z)\left((1 - a)\ln(1 - x^2) + 1\right) = 0.
\end{cases}
$$

From the first equation, we find $x = \pi n/2$, $n \in \mathbb{Z}$. Since the third equation contains the function $\ln(1 - x^2)$, the inequality $x^2 < 1$ is fulfilled, from which we get $n = 0$ and $x = y = 0$. The system takes the form

$$
\begin{cases}
z^2 = a - 1, \\
a\sin^2 z = 0.
\end{cases}
$$

For any of its solutions $(z; a)$, the pair $(-z; a)$ is also a solution of this system. Therefore, for uniqueness, it is necessary that z is equal to 0. Thus, if the system has a single solution, then it has the form $(0; 0; 0)$, and at the same time, $a = 1$. It remains to show that for $a = 1$, the system really has a unique solution.

2. Let $a = 1$. The system takes the form

$$
\begin{cases}
z\cos(x - y) + (2 + xy)\sin(x + y) - z = 0, \\
x^2 + y^2 + z^2 = 2(x + y), \\
x + y + \sin^2 z = 0.
\end{cases}
$$

Let us add up the second equation with the last one doubled. We get $x^2 + y^2 + z^2 + 2\sin^2 z = 0$, which gives $x = y = z = 0$. Therefore, we proved that for $a = 1$, the solution $(0; 0; 0)$ is the only one.

Answer: $a = 1$. □

Problems

6.1. For what values of b does the equation

$$\tan|b| = \log_2(\cos x - |x|)$$

have exactly one root?

6.2. Find all the values of a for which the following system has a unique solution:

$$\begin{cases} (|x| + 1)a = y + \cos x, \\ \sin^4 x + y^2 = 1. \end{cases}$$

6.3. Find all the values of a for which the equation

$$x^2 - 2a\sin(\cos x) + a^2 = 0$$

has a unique solution.

6.4. Find all the values of a for which the inequality

$$\cos 2x + a \leq 2\sqrt{x^2 + 16} - \frac{x^2 + 16}{a + \cos 2x}$$

has a unique solution.

6.5. Find all the values of a for which the equation

$$x^2 - |x - a + 6| = |x + a - 6| - (a - 6)^2$$

has a unique solution.

6.6. Find all the values of b for which the system of equations

$$\begin{cases} (x^2 + 1)b = y + \cos 2x, \\ 2^{|\sin x|} + |y| = 2 \end{cases}$$

has a unique solution.

6.7. Find all the values of the parameter a for which the equation

$$x^2 + (1 - a + \sqrt[4]{|x|})^2 = \frac{a^2}{4}$$

has exactly three solutions.

6.8. Find all the values of b for which the equation

$$b^2 x^2 - b\tan(\cos x) + 1 = 0$$

has a unique solution.

6.9. Find all the values of the parameter a and b for which the system of equations

$$\begin{cases} a + \sin bx \leq 1, \\ x^2 + ax + 1 \leq 0 \end{cases}$$

has a unique solution.

6.10. Find all the values of the parameters a and b for which the system of equations

$$\begin{cases} x^2 + y^2 = 2, \\ |y| - x = b \end{cases}$$

has exactly three solutions.

6.11. Find all the values of a for which the equation

$$\left| \frac{x(2^x - 1)}{2^x + 1} + 2a \right| = a^2 + 1$$

has an odd number of solutions.

6.12. Find all the values of the parameter a for which the system of equations

$$\begin{cases} x^4 - (a-1)\sqrt{a+3}\,y + a^4 + 2a^3 - 9a^2 - 2a + 8 = 0, \\ y = \sqrt{a+3}\,x^2 \end{cases}$$

has exactly three solutions.

6.13. Find all positive values of the parameter a for which the equation

$$\log_{2-x}(a^{2+x} + 2a^{1-x} + x - 1) + \log_{2+x}(a^{2-x} + 2a^{1+x} - x - 1) = 2$$

has exactly one solution (relative to x).

6.14. Find all the values of the parameter a for which the system of equations

$$\begin{cases} |bx| - |y| = 2a, \\ (x-b)^2 + y^2 = a^2 \end{cases}$$

has exactly three solutions.

6.15. Find all the values of the parameter a for which the system of equations

$$\begin{cases} 3 \cdot 2^{|x|} + 5|x| + 4 = 3y + 5x^2 + 3a, \\ x^2 + y^2 = 1 \end{cases}$$

has a unique solution.

6.16. Find all the values of the parameter a for which the system of equations

$$\begin{cases} (2 - \sqrt{3})^x + (2 + \sqrt{3})^x - 5 = a - 2y + y^2, \\ x^2 + (2 - a - a^2)y^2 = 0, \\ 0 \le y \le 2 \end{cases}$$

has a unique solution.

6.17. For what values of a and b does the system

$$\begin{cases} \dfrac{\tan^{-1} y}{x^2 + 1} \cdot \dfrac{x^y - 1}{x^y + 1} = a, \\ (y^2 - 1)^2 + b = x \end{cases}$$

have exactly five different solutions?

6.18. Find all the values of a for which the equation

$$\frac{8}{\pi} \tan^{-1}(1 + x/4) \cdot \log_{\sqrt{17}+4}(x + 4 + \sqrt{x^2 + 8x + 17})$$

$$= a^2 - a \sin\left(\pi \cdot \frac{x^2 + 8x - 64}{32}\right) - 2$$

has a unique solution, and determine this solution.

6.19. Find all the values of a for which the system

$$\begin{cases} (3\sqrt{x|x|} + |y| - 3)(|x| + 3|y| - 9) = 0, \\ (x - a)^2 + y^2 = 25 \end{cases}$$

has exactly three different solutions.

6.20. Find all the values of a for which the inequality

$$\sqrt[4]{x^2 - 6ax + 10a^2} + \sqrt[4]{3 + 6ax - x^2 - 10a^2}$$

$$\geq \sqrt[4]{C(a)} + |y - \sqrt{2}a^2| + |y - \sqrt{3}a|$$

has a unique solution. Here, $C(a) = \sqrt{3}a + 24 - \frac{3}{\sqrt{2}}$.

6.21. Find all the values of α for which the equation

$$x^2 + \frac{6x}{\sqrt{\sin \alpha}} + \frac{9\sqrt{3}}{\cos \alpha} + 36 = 0$$

has a unique solution.

6.22. Find all the values of a for which the equation

$$2\pi^2(x - 2)^2 + 4a \cos(2\pi x) - 25a^3 = 0$$

has a unique solution.

6.23. Find all the values of a for which the equation

$$|(x - 1)^2 - 2^{1-a}| + |x - 1| - (1 - x)^2 + 2^{a-1} = 4 + 4^a$$

has five different solutions.

6.24. Find all the values of a for which the equation

$$9^{-x+1} \cdot 3^{x^2} + a^3 + 5a^2 + a + \sqrt{2} = \sin(\pi x/4) + \cos(\pi x/4) + 3$$

has a unique solution.

6.25. Find all the values of a for which the system

$$\begin{cases} y - a^2 + 5(a - 1) = (a^2 - 5a + 6)(x - 3)^6 + \sqrt{(x - 3)^2}, \\ x^2 + y^2 = 2(3x - 4) \end{cases}$$

has exactly three solutions.

6.26. Find all the values of b for which the system of inequalities

$$\begin{cases} by^2 + 4by - 2x + 7b + 4 \leq 0, \\ bx^2 - 2y - 2bx + 4b - 2 \leq 0 \end{cases}$$

has a unique solution.

6.27. Find all the values of b for which the system

$$\begin{cases} b\sin|2z| + \log_5(x\sqrt[8]{2-5x^8}) + b^2 = 0, \\ ((y^2-1)\cos^2 z - y \cdot \sin 2z + 1) \cdot (1 + \sqrt{\pi + 2z} + \sqrt{\pi - 2z}) = 0 \end{cases}$$

has one or two solutions; determine these solutions.

6.28. Find all the values of a for which the system

$$\begin{cases} 1 - \sqrt{|x-1|} = \sqrt{7|y|}, \\ 49y^2 + x^2 + 4a = 2x - 1 \end{cases}$$

has exactly four different solutions.

6.29. Find all the values of b for which the equation

$$b^2 \sin\left(\frac{\pi+2}{2} - x\right) + \sin^2\left(\frac{2x}{b+1} - \frac{2}{b+1}\right)$$

$$- b\sqrt{4x^2 + 8 - 8x} = 3 + \arcsin|1 - x|$$

has a unique solution.

6.30. For what values of a does the equation

$$|x| + \left|\frac{2x-1}{3x-2}\right| = a$$

have exactly three different solutions?

6.31. Find all the values of b for each of which the system of inequalities

$$\begin{cases} y \geq (x-b)^2, \\ x \geq (y-b)^2 \end{cases}$$

has a unique solution.

6.32. Find all the values of a for each of which the following systems of equations are equivalent:

$$\begin{cases} x + 2y = 2 - a, \\ -x + ay = a - 2a^2 \end{cases}$$

and

$$\begin{cases} x^2 - y^4 - 4x + 3 = 0, \\ 2x^2 + y^2 + (a^2 + 2a - 11)x + 12 - 6a = 0. \end{cases}$$

6.33. Find all the values of a for which the following equation

$$2^{\frac{(x+1)^2}{x^2+1}} + a^2 - 4 = 2a\cos\left(\frac{x^2 - 1}{2x}\right)$$

has a unique solution.

6.34. Find all the values of a for which the equation

$$2^{\frac{2x}{1+x^2}} + a\cos\left(\frac{x^2 - 1}{x}\right) + a^2 - \frac{5}{4} = 0$$

has a unique solution.

6.35. Find all the values of a such that the equation

$$a^3\left(\tan^{-1}x - \tan^{-1}\frac{1}{x}\right)^2$$

$$= 4a + 5 - a^2 - \frac{2(a+1)}{\pi}\left(\arcsin\frac{2x}{x^2+1}\right)$$

has a unique solution.

6.36. Find all the values of a such that the system of inequalities

$$\begin{cases} 2(a+2) = (x-2)^2 + (y-2)^2 + z^2, \\ (xy + 4)\sin(x+y) + \cos(x-y) = 1, \\ \left(2 - \frac{xyz(a-2)}{\sqrt{1-2xy}}\right)(a\tan^2 z + x + y) = 0 \end{cases}$$

has a unique solution.

Hints and Answers

6.1. $b = \pi n$, $n \in \mathbb{Z}$.

6.2. $a = 2$.

6.3. $a = 0$, $a = 2\sin 1$.

6.4. $a = 3$.

6.5. $a = 4$, 8.

6.6. $b = 2$.

6.7. $a = 2$.

6.8. $b = \cot 1$.

6.9. $a = 2$, $b = \pi/2 + 2\pi n$, $n \in \mathbb{Z}$; $a = -2$, $b \in \mathbb{R}$.

6.10. $b = \sqrt{2}$.

6.11. $a = \pm 1$.

6.12. $a = 2$.

6.13. $a = 1$.

6.14. $(a; b) = (t^2/(|t|+2); t)$, where $t \neq 0$; or $(a; b) = (t^2/(|t|-2); t)$, where $|t| > 2$.

6.15. $a = 4/3$.

6.16. $a = -3$, $a = -2$.

6.17. $a = 0$, $b \in (0; 1)$.

6.18. When $a = 1$, $x = -4$.

6.19. $a = -4$, 4, 6.

6.20. $a = \sqrt{3/2}$.

6.21. $a = 5\pi/6 + 2\pi l$, $\pi/18 + 2\pi m$, $13\pi/18 + 2\pi n$, $l, m, n \in \mathbb{Z}$.

6.22. $a = 0$, $a = -2/5$.

6.23. $a = -1$.

6.24. $a = 0$, $a = (-5 \pm \sqrt{21})/2$.

6.25. $a = 2$, $a = 3$.

6.26. $b = 1/3$.

6.27. If $b = -1/2\sqrt{2}$, then one solution is $(1/\sqrt[8]{5}; 0; 0)$; if $b = -1/2 + \sqrt{3/8}$, then two solutions are $(1/\sqrt[8]{5}; 1; \pi/4)$ and $(1/\sqrt[8]{5}; -1; -\pi/4)$.

6.28. $a = -1/32$, $a = -1/4$.

6.29. $b = 3$.

6.30. $a = 2/3$, $a = 2$.

6.31. $b = -1/4$.

6.32. $a = -2$, $a = -1$. **Hint:** Solve the first equation in the system and then use symmetries to investigate the second equation in the system.

6.33. $a = 0$, $a = 3$.

6.34. $a = -3/2$.

6.35. $a = 6$. **Hint:** The equations $f(x) = 0$ and $f(1/x) = 0$ are equivalent.

6.36. $a = 2$.

Chapter 7

Problems That Reduce to the Study of a Quadratic Equation and the Selecting of Non-Negative Expressions

7.1 Problems that Reduce to the Study of a Quadratic Equation

For the quadratic equation

$$ax^2 + bx + c = 0, \quad a \neq 0, \tag{7.1}$$

we distinguish three cases:

1. If $D = b^2 - 4ac < 0$, then there are no real solutions to quadratic equation (7.1) (see Figs. 7.1 and 7.3).
2. If $D = b^2 - 4ac = 0$, then the solution of quadratic equation (7.1) takes the form $x = -b/2a$ (see Figs. 7.2 and 7.4).
3. If $D = b^2 - 4ac > 0$, then quadratic equation (7.1) has two roots x_+, x_- (see Figs. 7.5 and 7.6):

$$x_\pm = \frac{-b \pm \sqrt{b^2 - 4ac}}{2a}.$$

In addition, the equality $ax^2 + bx + c = a(x - x_+)(x - x_-)$ is fulfilled.

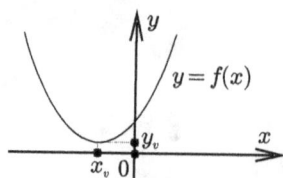

Fig. 7.1. $D < 0$, $a > 0$.

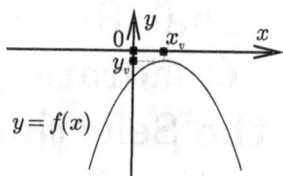

Fig. 7.2. $D = 0$, $a > 0$.

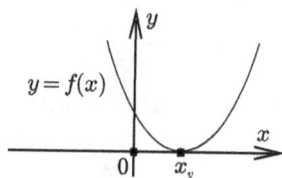

Fig. 7.3. $D < 0$, $a < 0$.

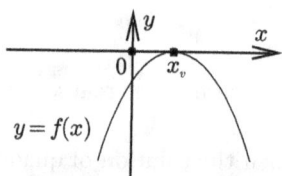

Fig. 7.4. $D = 0$, $a < 0$.

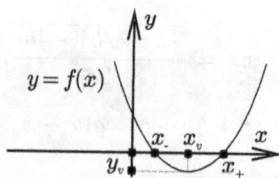

Fig. 7.5. $D > 0$, $a < 0$.

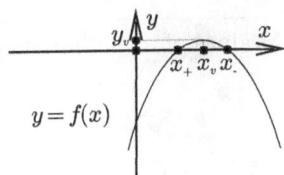

Fig. 7.6. $D > 0$, $a > 0$.

1. An important role in solving quadratic equations with a parameter is played by **Vieta's theorem**. For the quadratic equation $ax^2 + bx + c = 0$, $a \neq 0$, with roots $x_\pm$ (the case of $D \geq 0$), Vieta's formulas are executed:

$$x_+ + x_- = -\frac{b}{a};$$

$$x_+ x_- = \frac{c}{a}.$$

2. The second important point is that when solving problems that reduce to the study of quadratic equations, one must remember the geometric interpretation of the quadratic equation. For example, selecting a full square, we get (for $a \neq 0$)

$$ax^2 + bx + c = a \cdot \left(x + \frac{b}{2a}\right)^2 + \left(c - \frac{b^2}{4a}\right)$$

$$= a \cdot (x - x_v)^2 + y_v,$$

where

$$x_v = -\frac{b}{2a}, \quad y_v = c - \frac{b^2}{4a}.$$

The graph of the function $y = ax^2 + bx + c$ is a parabola whose vertex has coordinates $(x_v; y_v)$. When $a > 0$, the branches of the parabola are directed upward, and the minimum of the quadratic function is reached at the vertex of the parabola. When $a < 0$, the branches of the parabola are directed downward, and the maximum of the quadratic function is reached at the vertex of the parabola.

Example 7.1. For what values of a the function

$$y = \frac{2^{ax+7}}{2^{x^2}}$$

has a maximum at the point $x = 4$?

Proof. The original function is represented as

$$y = 2^{-x^2 + ax + 7}.$$

Since the function 2^t monotonically increases, the maximum of the function $y = 2^{-x^2+ax+7}$ is reached at the same point as the quadratic function $f(x) = -x^2 + ax + 7$. The branches of this parabola are directed downward. Therefore, the maximum is reached at the vertex of the parabola, i.e., at the point $x_v = a/2$. However, according to the condition, $x_v = 4$, hence $a = 8$.

Answer: $a = 8$. $\qquad\qquad\qquad\qquad\qquad\qquad\qquad\qquad\qquad\qquad\square$

Example 7.2. For each value of a, solve the inequality

$$ax^2 + x + 3a^3 > 0.$$

Proof. Let $a = 0$. Then, the solution to the inequality will be the set of numbers $x > 0$.

For $a \neq 0$, the function $f(x) = ax^2 + x + 3a^3$ is quadratic, whose graph is a parabola. Consider three cases depending on the sign of the discriminant $D = 1 - 12a^4$ of the function $f(x)$, i.e., $D < 0$, $D = 0$, and $D > 0$:

1. Let $D = 1 - 12a^4 < 0$, i.e., $a \in (-\infty; -1/\sqrt[4]{12}) \cup (1/\sqrt[4]{12}; +\infty)$. Then, depending on the sign of a, the function $f(x)$ will be positive or negative everywhere (see Figs. 7.7 and 7.8).

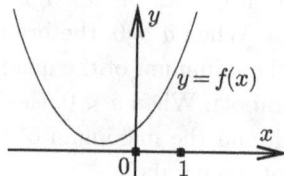

Fig. 7.7. $D < 0$, $a > 0$.

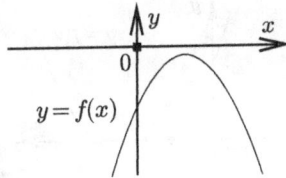

Fig. 7.8. $D < 0$, $a < 0$.

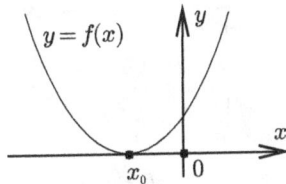

Fig. 7.9. $D = 0$, $a > 0$.

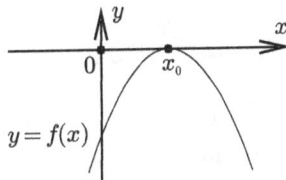

Fig. 7.10. $D = 0$, $a < 0$.

For $a > 1/\sqrt[4]{12}$, we get $f(x) > 0$, for any $x \in \mathbb{R}$. For $a < -1/\sqrt[4]{12}$ we get $f(x) < 0$, for any $x \in \mathbb{R}$.

Partial answer: If $a < -1/\sqrt[4]{12}$, there are no solutions; if $a > 1/\sqrt[4]{12}$, then $x \in (-\infty; +\infty)$.

2. Let $D = 1 - 12a^4 = 0$, i.e., $a = \pm 1/\sqrt[4]{12}$. Then, the quadratic equation $f(x) = 0$ will have a single root $x_0 = -1/2a$ (see Figs. 7.9 and 7.10). For $a = 1/\sqrt[4]{12}$, we get $f(x) > 0$, for any $x \in \mathbb{R}\backslash\{-\sqrt[4]{12}/2\}$. For $a = -1/\sqrt[4]{12}$, we get $f(x) \leq 0$, for any $x \in \mathbb{R}$.

Partial answer: If $a = -1/\sqrt[4]{12}$, there are no solutions; if $a = 1/\sqrt[4]{12}$, then $x \in (-\infty; +\infty)\backslash\{-\sqrt[4]{12}/2\}$.

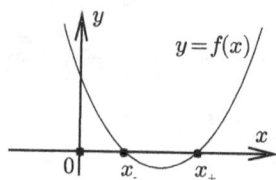

Fig. 7.11. $D > 0$, $a > 0$.

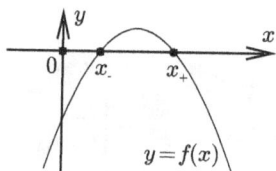

Fig. 7.12. $D < 0$, $a < 0$.

3. Let $D = 1 - 12a^4 > 0$, i.e., $a \in (-1/\sqrt[4]{12}; 1/\sqrt[4]{12})$. Then, the quadratic equation $f(x) = 0$ will have two solutions (see Figs. 7.11 and 7.12).

$$x_+ = \frac{-1 + \sqrt{1 - 12a^4}}{2a}, \qquad x_- = \frac{-1 - \sqrt{1 - 12a^4}}{2a}.$$

For $a > 0$, we get $f(x) > 0$, for any $x \in (-\infty; x_-) \bigcup (x_+; +\infty)$. For $a < 0$, we get $f(x) > 0$, for any $x \in (x_+; x_-)$.

Partial answer: If $0 < a < 1/\sqrt[4]{12}$, then $x \in (-\infty; (-1 - \sqrt{1 - 12a^4})/2a) \bigcup ((-1 + \sqrt{1 - 12a^4})/2a; +\infty)$; if $-1/\sqrt[4]{12} < a < 0$, then $x \in ((-1 + \sqrt{1 - 12a^4})/2a; (-1 - \sqrt{1 - 12a^4})/2a)$.

Combining the partial answers, we get the answer.

Answer: If $a \leq -1/\sqrt[4]{12}$, there are no solutions; if $-1/\sqrt[4]{12} < a < 0$, then $x \in ((-1 + \sqrt{1 - 12a^4})/2a; (-1 - \sqrt{1 - 12a^4})/2a)$; if $a = 0$, then $x \in (0; +\infty)$; if $0 < a \leq 1/\sqrt[4]{12}$, then $x \in (-\infty; (-1 - \sqrt{1 - 12a^4})/2a) \bigcup ((-1 + \sqrt{1 - 12a^4})/2a; +\infty)$; and if $a > 1/\sqrt[4]{12}$, then $x \in (-\infty; +\infty)$.   □

Example 7.3. Find all values of b for which the equation

$$x - 2 = \sqrt{2(b - 1)x + 1}$$

has a unique solution.

Proof. Transform the equation:

$$
\begin{cases} x \geq 2, \\ (x-2)^2 = 2(b-1)x + 1. \end{cases} \iff \begin{cases} x \geq 2, \\ x^2 - 2(b+1)x + 3 = 0. \end{cases}
$$

The parabola $f(x) = x^2 - 2(b+1)x + 3$ has branches which are directed upward, so the a solution is possible only in the following cases (see the corresponding Figs. 7.13–7.18):

$$
\text{(I)} \begin{cases} D > 0, \\ f(2) < 0; \end{cases} \quad \text{(II)} \begin{cases} D > 0, \\ f(2) = 0, \\ x_v < 2; \end{cases} \quad \text{(III)} \begin{cases} D = 0, \\ x_v \geq 2. \end{cases}
$$

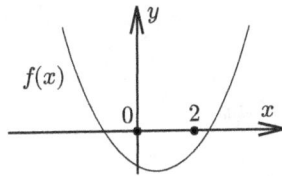

Fig. 7.13.   Case I.

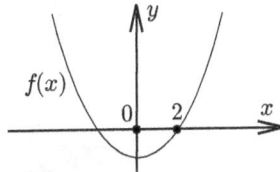

Fig. 7.14.   Case I.

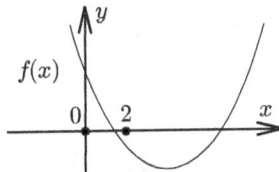

Fig. 7.15.   Case II.

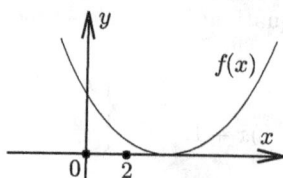

Fig. 7.16.   Case III.

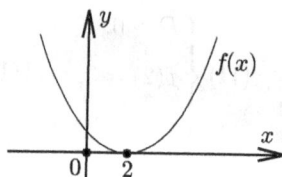

Fig. 7.17.   The case of two roots.

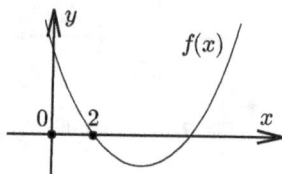

Fig. 7.18.   Case III.

Find the discriminant of the equation $f(x) = 0$:

$$D = 4(b+1)^2 - 4 \cdot 3 = 4(b+1-\sqrt{3})(b+1+\sqrt{3}).$$

Let us now analyze each of the three cases listed above:

$$\text{Case I}: \begin{cases} D > 0, \\ f(2) < 0. \end{cases} \iff \begin{cases} b \in (-\infty; -1-\sqrt{3}) \cup (-1+\sqrt{3}; +\infty), \\ 4 - 4(b+1) + 3 < 0. \end{cases}$$

$$\iff \begin{cases} b \in (-\infty; -1-\sqrt{3}) \cup (-1+\sqrt{3}; +\infty), \\ b > 3/4. \end{cases}$$

Compare the numbers $-1 + \sqrt{3}$ and $3/4$:

$$-1 + \sqrt{3} \bigvee 3/4,$$

$$\sqrt{3} \bigvee 7/4,$$

$$4\sqrt{3} \bigvee 7,$$

$$48 < 49.$$

Thus, in the first case, we get $b > 3/4$. Let us analyze the second case. (The second case has to be dealt separately from the first since a situation is possible (see Fig. 7.17) where $D > 0$ and $f(2) = 0$, but at the same time we have two solutions (the case of $x_v > 2$).)

$$\text{Case II} : \begin{cases} D > 0, \\ f(2) = 0, \\ x_v < 2. \end{cases} \Longleftrightarrow \begin{cases} b \in (-\infty; -1 - \sqrt{3}) \cup (-1 + \sqrt{3}; +\infty), \\ b = 3/4, \\ \dfrac{2(b+1)}{2} < 2. \end{cases}$$

$$\Longleftrightarrow b = 3/4.$$

It remains to deal with the third and final case:

$$\text{Case III} : \begin{cases} D = 0, \\ x_v \geq 2. \end{cases} \Longleftrightarrow \begin{cases} b = -1 \pm \sqrt{3}, \\ b \geq 1. \end{cases} \Longleftrightarrow b \in \emptyset.$$

Combining the results of these cases, we get the answer.

Answer: $b \in [3/4; +\infty)$. $\square$

Example 7.4. Find all the values of a for each of which the equation

$$a 4^{\frac{1}{x}-1} + (a-1)2^{\frac{1}{x}} - a^3 + 3a - 2 = 0$$

has no roots.

Proof. With respect to the variable $t = 2^{\frac{1}{x}-1}$, the equation takes the form

$$at^2 + 2(a-1)t - a^3 + 3a - 2 = 0.$$

The function $f(x) = 2^{\frac{1}{x}-1}$ accepts all positive values except $\frac{1}{2}$. Therefore, the original equation has no roots if the roots of the equation with respect to t do not belong to the set $(0; \frac{1}{2}) \cup (\frac{1}{2}; +\infty)$. For $a = 0$, we have $t = -1$, so this a is suitable. For $a \neq 0$, this is a quadratic equation whose roots are $t_1 = a - 1$, $t_2 = -\frac{(a-1)(a+2)}{a}$. We find the set of values of a for which $t_1, t_2 \in (-\infty; 0] \cup \{\frac{1}{2}\}$:

$$
\left\{
\begin{array}{l}
\left[
\begin{array}{l}
a - 1 \leqslant 0, \\
a - 1 = \dfrac{1}{2}.
\end{array}
\right. \\[2em]
\left[
\begin{array}{l}
-\dfrac{(a-1)(a+2)}{a} \leqslant 0, \\[1em]
-\dfrac{(a-1)(a+2)}{a} = \dfrac{1}{2}.
\end{array}
\right.
\end{array}
\right.
\Longleftrightarrow
\left\{
\begin{array}{l}
\left[
\begin{array}{l}
a \leqslant 1, \\
a = \dfrac{3}{2},
\end{array}
\right. \\[2em]
\left[
\begin{array}{l}
a \geqslant 1, \\
-2 \leqslant a < 0, \\
a = \dfrac{-3 \pm \sqrt{41}}{4}.
\end{array}
\right.
\end{array}
\right.
$$

$$
\Longleftrightarrow
\left[
\begin{array}{l}
-2 \leqslant a < 0, \\
a = 1, \\
a = \dfrac{3}{2}, \\[1em]
a = \dfrac{-3 \pm \sqrt{41}}{4}.
\end{array}
\right.
$$

Answer: $-2 \leqslant a \leqslant 0$, $a = 1$; $\frac{3}{2}$; $\frac{-3 \pm \sqrt{41}}{4}$.   □

Example 7.5. Find all the values of a such that the inequality

$$(x + \log_4 |a|)(x + \log_{|a|} 4)(x^2 + 10 \cdot 2^a x + a^2 - 3) \geq 0$$

holds for any x.

Proof. Let $g(x) = (x + \log_4 |a|)(x + \log_{|a|} 4)$, $h(x) = x^2 + 10 \cdot 2^a x + a^2 - 3$, and $d_h(a)$ be a quarter of the discriminant of the square trinomial $h(x)$. The inequality holds for all x if and only if:

1. the roots of $g(x)$ coincide, and the roots of $h(x)$ are either absent or coincide;
2. a pair of roots $g(x)$ coincides with a pair of roots $h(x)$.

The conditions in point (a) mean that $\log_4 |a| = \log_{|a|} 4$, $d_h(a) \leq 0$. Solving the equation, we get $\log_4 |a| = \log_{|a|} 4 \Leftrightarrow (\log_4 |a|)^2 = 1$.

If $\log_4 |a| = 1 \Leftrightarrow |a| = 4$, then either $a = 4$ and $d_h(4) = (5 \cdot 16)^2 - 13 > 0$
or $a = -4$ and $d_h(-4) = (5/16)^2 - 13 < 0$.

If $\log_4 |a| = -1 \Leftrightarrow |a| = \frac{1}{4}$, then the constant term $h(x)$ is $a^2 - 3 = \frac{1}{16} - 3 = -\frac{47}{16} < 0$; therefore, $h(x)$ has two different roots.

The conditions in point (b) mean that $h(x) = g(x)$. Therefore,

$$\begin{cases} a^2 - 3 = \log_4 |a| \cdot \log_{|a|} 4 \equiv 1, \\ 10 \cdot 2^a = \log_4 |a| + \log_{|a|} 4. \end{cases}$$

The first equation of the system means that $a^2 - 3 = 1 \Leftrightarrow |a| = 2$, and then

$$\log_4 |a| + \log_{|a|} 4 = 10 \cdot 2^a = \frac{5}{2}.$$

The last equality holds only when $a = -2$.

Answer: $a = -4$, $a = -2$.   □

Example 7.6. Find all the values of a such that the equation

$$(x^2 - 2(a+1)x + a^2 + 2a)^2 + (a+5)(x^2 - 2(a+1)x + a^2 + 2a)$$

$$- a^2 - 7a - 10 = 0$$

has (a) a unique solution and (b) exactly two different solutions.

Proof. Replacing a variable

$$y = x^2 - 2(a+1)x + a^2 + 2a$$

(the graphs of the function $y = f(x)$ are shown in Fig. 7.19) reduces the
original equation to the quadratic equation $g(y) = 0$, where

$$g(y) = y^2 + (a+5)y - a^2 - 7a - 10.$$

If y_0 is the root of this equation, then to find the roots of the original
equation, you need to solve the equation

$$x^2 - 2(a+1)x + a^2 + 2a = y_0,$$

or the equivalent equation

$$(x - (a+1))^2 = y_0 + 1.$$

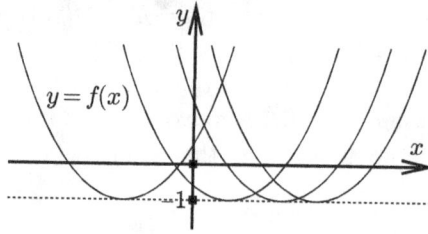

Fig. 7.19.   Graph of $f(x)$ with different a.

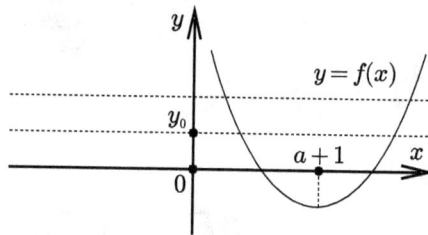

Fig. 7.20.   Roots of $f(x) = y_0$.

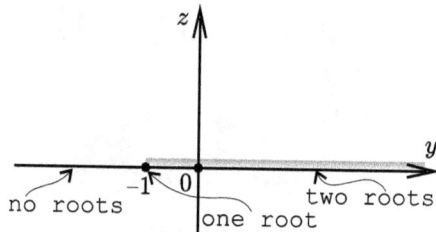

Fig. 7.21.   Number of roots.

- If $y_0 < -1$, this equation has no roots since its left part is non-negative for any x and the right part is negative.
- If $y_0 = -1$, then the equation has one root $x = a + 1$.
- If $y_0 > -1$, then the equation has two roots, one of which is less than $a + 1$, and the other is greater (see Fig. 7.20).

We depict these remarks on a graph (see Fig. 7.21). Consider point (a) of the original problem. According to the above, the original equation has one

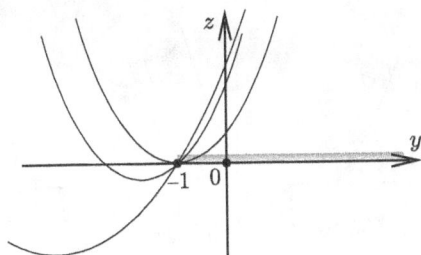

Fig. 7.22.   Case $g(-1) = 0$.

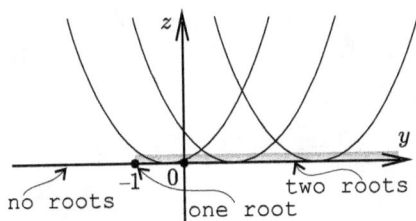

Fig. 7.23.   Number of roots.

root if and only if the equation $g(y) = 0$ either has a root $y_0 = -1$ of multiplicity 2 or, besides the root $y_0 = -1$, has a root less than -1. In both cases, the abscissa of the vertex of the parabola is less than or equal to -1 (see Fig. 7.22). These conditions can be combined into a system:

$$\begin{cases} g(-1) = 0, \\ y_v \leq -1. \end{cases} \iff \begin{cases} -a^2 - 8a - 14 = 0, \\ -(a+5)/2 \leq -1. \end{cases} \iff a = -4 + \sqrt{2}.$$

Consider point (b) of the problem. The original equation has two solutions in one of two cases. In the first of them, the equation $g(y) = 0$ has a single root $y_0 > -1$, which coincides with y_v (see Fig. 7.23). This case corresponds to the system

$$\begin{cases} D = 0, \\ y_v > -1. \end{cases} \iff \begin{cases} (a+5)^2 + 4(a^2 + 7a + 10) = 0, \\ -(a+5)/2 > -1. \end{cases}$$

$$\iff \begin{cases} 5a^2 + 38a + 65 = 0, \\ -(a+5)/2 > -1. \end{cases} \iff a = -5.$$

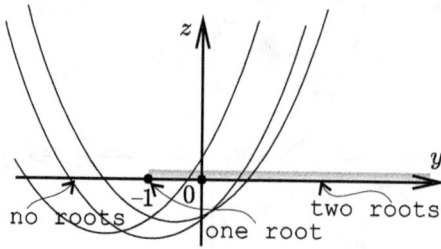

Fig. 7.24.   Number of roots.

In the second case (see Fig. 7.24), the equation $g(y) = 0$ has two roots, one of which is larger than -1, and the other is less than -1, which is equivalent to the condition $g(-1) < 0$, i.e.,

$$g(-1) < 0 \iff a^2 + 8a + 14 > 0 \iff a \in (-\infty; -4 - \sqrt{2})$$
$$\cup(-4 + \sqrt{2}; +\infty).$$

Answer: (a) There is only one solution for $a = -4 + \sqrt{2}$; (b) there are exactly two different solutions for $a \in (-\infty; -4 - \sqrt{2}) \cup \{-5\} \cup (-4 + \sqrt{2}; +\infty)$. $\square$

7.2   Selecting Full Squares and Non-Negative Expressions

When solving these problems, it is necessary to remember the squaring formulas for several terms:

$(a + b)^2 = a^2 + b^2 + 2ab,$

$(a + b + c)^2 = a^2 + b^2 + c^2 + 2ab + 2ac + 2bc,$

$(a + b + c + d)^2 = a^2 + b^2 + c^2 + d^2 + 2ab + 2ac + 2ad + 2bc + 2bd + 2cd.$

It is easy to note (and prove by mathematical induction) that when squaring the expression $a_1 + a_2 + \cdots + a_n$, we end up with the sum of all squares a_k^2, $k = 1, 2, \ldots n$ and all possible pairwise products $2a_k a_l$, $1 \leq k < l \leq n$, $k, l \in \mathbb{Z}$, i.e., the following general formula holds:

$$(a_1 + a_2 + \cdots + a_n)^2 = \sum_{k=1}^{n} a_k^2 + 2 \cdot \sum_{1 \leq k < l \leq n} a_k a_l.$$

Example 7.7. Find all the values of the parameter a for each of which the system

$$\begin{cases} x^2 + 2y^2 + 2y(x-a) + a^2 = 0, \\ 2^{-2-y} \cdot \log_2 x < 1 \end{cases}$$

has solutions, and find those solutions.

Proof. The first equation is reduced to the form $(x+y)^2 + (y-a)^2 = 0$, from which we get $x = -a$ and $y = a$. Substituting into the inequality gives

$$2^{-2-a} \cdot \log_2(-a) < 1 \Longleftrightarrow \log_2(-a) < 2^{2+a}.$$

Equality is achieved when $a = -2$, and due to monotony, we get $a \in (-2; 0)$.

Answer: $x = -a$ and $y = a$ if $a \in (-2; 0)$. ☐

Example 7.8. Find all values of a for which the equation

$$(x^2 - 6|x| + a)^2 + 10(x^2 - 6|x| + a) + 26 = \cos(16\pi/a)$$

has exactly two distinct roots.

Proof. The original equation is equivalent to the equation

$$(x^2 - 6|x| + a)^2 + 2\cdot 5(x^2 - 6|x| + a) + 25 + 1 - \cos(16\pi/a) = 0$$
$$\Longleftrightarrow (x^2 - 6|x| + a + 5)^2 + (1 - \cos(16\pi/a)) = 0.$$

The function $(x^2 - 6|x| + a + 5)^2$ and the value $1 - \cos(16\pi/a)$ are non-negative for all values of variables. We have obtained that the sum of the non-negative terms is zero. This is the case if and only if these terms vanish simultaneously, i.e., the original equation is equivalent to the system

$$\begin{cases} x^2 - 6|x| + a + 5 = 0, \\ 1 - \cos(16\pi/a) = 0. \end{cases} \iff \begin{cases} (|x| - 3)^2 = 4 - a, \\ a = 8/n, \quad n \in \mathbb{Z}. \end{cases}$$

$$\iff \begin{cases} a \le 4, \\ |x| = 3 \pm \sqrt{4 - a}, \\ a = 8/n, \quad n \in \mathbb{Z}. \end{cases}$$

The set of equations $|x| = 3 \pm \sqrt{4 - a}$ has exactly two roots if and only if either $\sqrt{4 - a} = 0$ or $3 - \sqrt{4 - a} < 0$, i.e., either $a = 4$ or $a < -5$. Therefore, the values of a must belong to the set $(-\infty; -5) \cup \{4\}$, from where, taking into account the condition

$$a = 8/n, \quad n \in \mathbb{Z} \iff a = \pm 8, \pm 4, \pm 8/3, \dots,$$

we find two values, $a = 4$ and $a = -8$, belonging to the set $(-\infty; -5) \cup \{4\}$.

Answer: $a = 4$, $a = -8$.   □

Example 7.9. For each value of a, solve the equation

$$9a^2 + \log_2^2 x + 3\arccos(x - 1) - (3a - 1)\log_2 x^2 - 6a + 1 = 0.$$

Proof. Transform the original equation:

$$9a^2 + \log_2^2 x + 3\arccos(x - 1) - (3a - 1)\log_2 x^2 - 6a + 1 = 0$$

$$\iff \log_2^2 x - 2 \cdot (3a - 1)\log_2 x + 9a^2 - 6a + 1 + 3\arccos(x - 1) = 0$$

$$\iff \log_2^2 x - 2 \cdot (3a - 1)\log_2 x + (3a - 1)^2 + 3\arccos(x - 1) = 0$$

$$\iff (\log_2 x - 3a + 1)^2 + 3\arccos(x - 1) = 0.$$

Since the functions $(\log_2 x - 3a + 1)^2$ and $3\arccos(x - 1)$ are non-negative, the original equation is equivalent to the system

$$\begin{cases} \log_2 x - 3a + 1 = 0, \\ 3\arccos(x - 1) = 0. \end{cases} \iff \begin{cases} a = 2/3, \\ x = 2. \end{cases}$$

Answer: If $a = 2/3$, then $x = 2$; for other a, there are no solutions.   □

Example 7.10. Find the sum of the first hundred positive roots of the equation

$$\cos(8\pi x) + 2\cos(4\pi x) - \cos(2\pi x) + 2\sin(\pi x) + 3 = 0.$$

Proof. Denote $\alpha = \pi x$.

$$\cos(8\alpha) + 2\cos(4\alpha) - \cos(2\alpha) + 2\sin(\alpha) + 3 = 0$$

$$\Longleftrightarrow 2\cos^2 4\alpha - 1 + 2\cos 4\alpha - 1 + 2\sin^2 \alpha + 2\sin \alpha + 3 = 0$$

$$\Longleftrightarrow 2\left(\cos 4\alpha + \frac{1}{2}\right)^2 + 2\left(\sin \alpha + \frac{1}{2}\right)^2 = 0.$$

We get

$$\begin{cases} \cos 4\alpha = -1/2, \\ \sin \alpha = -1/2. \end{cases} \Longleftrightarrow \begin{bmatrix} \alpha = -\dfrac{\pi}{6} + 2\pi m, & m \in \mathbb{Z}, \\ \\ \alpha = -\dfrac{5\pi}{6} + 2\pi n, & n \in \mathbb{Z}. \end{bmatrix}$$

Hence, $x = -1/6 + 2m$, $x = -5/6 + 2n$, $m, n \in \mathbb{Z}$. The sum of the first two positive roots is $1\frac{1}{6} + 1\frac{5}{6} = 3$. The next two positive roots are greater by 4, i.e., equal to 7, and so on. The required amount is equal to

$$3 + 7 + 11 + \cdots + (3 + 4 \cdot 49) = \frac{6 + 4 \cdot 49}{2} \cdot 50 = 5050.$$

Answer: 5050. □

Example 7.11. For each value of a, solve the equation

$$\sin^2 x + \sin^2 2x + \sin^2 3x - 2a(\sin x + \sin 2x + \sin 3x)$$

$$+ \cos x - \cos 3x + 2a^2 = 0.$$

Proof. Here are two solutions to this example:

1. Convert $\cos x - \cos 3x$ to the product of sines $\cos x - \cos 3x = 2\sin x \sin 2x$ and group the terms[1]:

$$\left(\sin^2 x + \sin^2 2x + a^2 - 2a(\sin x + \sin 2x) + 2\sin x \sin 2x\right)$$
$$+ \left(a^2 + \sin^2 3x - 2a\sin 3x\right) = 0$$
$$\Longleftrightarrow (\sin x + \sin 2x - a)^2 + (\sin 3x - a)^2 = 0.$$

However, since the sum of squares is a non-negative number, every term that is a complete square is zero, i.e.,

$$\begin{cases} \sin x + \sin 2x = a, \\ \sin 3x - a = 0. \end{cases} \Longleftrightarrow \begin{cases} \sin x + \sin 2x = \sin 3x, \\ a = \sin 3x. \end{cases}$$

$$\Longleftrightarrow \begin{cases} 2\sin(3x/2)\cos(x/2) = 2\sin(3x/2)\cos(3x/2), \\ a = \sin 3x. \end{cases}$$

$$\Longleftrightarrow \begin{cases} \sin(3x/2) \cdot \sin x \cdot \sin(x/2) = 0, \\ a = \sin 3x. \end{cases}$$

Solve the equation $\sin(3x/2)\sin x \sin(x/2) = 0$. Since the equality $\sin(x/2) = 0$ implies that $\sin x = 0$, we obtain the equivalent equation $\sin(3x/2)\sin x = 0$. On solving it, we find $x = \pi m$, $x = 2\pi n/3$, $m, n \in \mathbb{Z}$. It follows from the equality $a = \sin 3x$ that the equality $a = 0$ is satisfied for all the found x.

We get the following answer: if $a = 0$, then $x = \pi m$, $2\pi n/3$, $m, n \in \mathbb{Z}$; there are no solutions for $a \neq 0$.

2. Here is the second solution to the original equation. The equation can be considered a square with respect to a. Using the cosine difference formula $\cos x - \cos 3x = 2\sin x \sin 2x$, we get

$$2a^2 - 2a(\sin x + \sin 2x + \sin 3x) + \sin^2 x + \sin^2 2x + \sin^2 3x$$
$$+ 2\sin x \sin 2x = 0.$$

---

[1]Here, we use the formula of the square of the sum of three terms: $(a + b + c)^2 = a^2 + b^2 + c^2 + 2ab + 2ac + 2bc$.

Find the discriminant of this equation:

$$D/4 = (\sin x + \sin 2x + \sin 3x)^2$$
$$- 2(\sin^2 x + \sin^2 2x + \sin^2 3x + 2\sin x \sin 2x)$$
$$= -\sin^2 x - \sin^2 2x - \sin^2 3x - 2\sin x \sin 2x + 2\sin x \sin 3x$$
$$+ 2\sin 2x \sin 3x = -(\sin x + \sin 2x - \sin 3x)^2 \le 0.$$

But the solution of the quadratic equation exists only in the case of $D \ge 0$, so $D = -4(\sin x + \sin 2x - \sin 3x)^2 = 0$, from where, again, we arrive at the conclusion that if the equation has a solution only in the case of $\sin x + \sin 2x = \sin 3x$, and at the same time,

$$a = (\sin x + \sin 2x + \sin 3x)/2 = \sin 3x.$$

Thus, we arrive at the system again,

$$\begin{cases} \sin x + \sin 2x = \sin 3x, \\ a = \sin 3x, \end{cases}$$

on solved which we get the answer.

Answer: If $a = 0$, then $x = \pi m$, $2\pi n/3$, $m, n \in \mathbb{Z}$; there are no solutions for $a \ne 0$. ☐

Problems

7.1. For what values of a does the function

$$y = \frac{3^{x^2}}{3^{ax-11}}$$

have a minimum when $x = 6$?

7.2. For what values of a one of the roots of the equation

$$(a^2 + a + 1)x^2 + (2a - 3)x + a - 5 = 0$$

is greater than 1 and the other is less than 1?

7.3. Find all values of a such that the equation

$$ax^2 + (4a^2 - 3)x - 10 = 0$$

has two distinct roots whose moduli are equal.

7.4. Find all the values of a for which the inequality

$$x^2 + ax + a^2 + 6a < 0$$

holds for all $x \in (1; 2)$.

7.5. For what values of a does the equation

$$\frac{(a + 4)x^2 + 6x - 1}{x + 3} = 0$$

have a unique solution?

7.6. Find all the values of a for which the inequality

$$ax^2 - 4x + 3a + 1 < 0$$

holds for all $x > 0$.

7.7. Find all the values of a for each of which, among the roots of the equation

$$ax^2 + (a + 4)x + a + 1 = 0,$$

there is exactly one negative root.

7.8. Find all the values of a for which, from the inequality

$$x^2 - (3a + 1)x + a > 0,$$

it follows that $x > 1$.

7.9. Find all the values of a such that the smallest value of the function

$$f(x) = 4x^2 + 4ax + a^2 - 2a + 2$$

on the set $1 \leq |x| \leq 3$ is at least 6.

7.10. For what values of a does the system

$$\begin{cases} x^4 - (a-1)\sqrt{a+3}\,y + a^4 + 2a^3 - 9a^2 - 2a + 8 = 0, \\ y = \sqrt{a+3}\,x^2 \end{cases}$$

have exactly three different solutions?

7.11. For what values of a does the equation

$$(a-1) \cdot 4^x + (2a-3) \cdot 6^x = (3a-4) \cdot 9^x$$

have a single solution?

7.12. Find all the values of a for which the equation

$$4^x + (a^2 + 5) \cdot 2^x + 9 - a^2 = 0$$

has no solutions.

7.13. Find all the values of a for each of which the system

$$\begin{cases} -x^2 + 12x - a \geq 0, \\ x \leq 2 \end{cases}$$

holds with at least one value of x.

7.14. For what values of a does the inequality

$$3 \cdot 4^x - 6a \cdot 2^x + 3a^2 + 2a - 14 < 0$$

have no solutions?

7.15. Find all the values of a for which the equation

$$(a+1)4^{\frac{1}{x}+2} + a2^{\frac{1}{x}+3} - a^3 - 3a^2 = 0$$

has at least one root.

7.16. For what values of a does the equation

$$2a(x+1)^2 - |x+1| + 1 = 0$$

have four different roots?

7.17. Find all the values of a for which the inequality

$$\log_{1/5}(x^2 - ax + 7) < -1$$

is executed for all values of x from the interval $x < 0$.

7.18. Find all integer values of a such that the equation

$$\sqrt[3]{x^6} - \left(\frac{1}{a} - 2\right) \cdot \sqrt[4]{x^4} + 1 - \frac{2}{a} = 0$$

has solutions and they are all integers.

7.19. Denote by x_1 and x_2 the roots of the square trinomial

$$(a - 1)x^2 - (2a + 1)x + 2 + 5a.$$

1. Find all the values of a for which $x_1 > 1$ and $x_2 > 1$.
2. Find all the values of b for which the function $(x_1 - b)(x_2 - b)$ takes a constant value for all a at which it is defined.

7.20. Find all the values of a for which the sum of the arctangents of the roots of the equation

$$x^2 + (1 - 2a)x + a - 4 = 0$$

are greater than $\pi/4$.

7.21. Find all the values of p for which the equation

$$x - 2 = \sqrt{-2(p + 2)x + 2}$$

has a unique solution.

7.22. For each value of a, find all solutions to the inequality

$$x + 2a - 2\sqrt{3ax + a^2} > 0.$$

7.23. Find all the values of b for which the equation

$$3 \cdot \sqrt[5]{x + 2} - 16b^2 \cdot \sqrt[5]{32x + 32} = \sqrt[10]{x^2 + 3x + 2}$$

has a unique solution.

7.24. Find all the values of a for each of which the equation

$$(a - 1)\cos^2 x - (a^2 + a - 2)\cos x + 2a^2 - 4a + 2 = 0$$

has more than one solution on the segment $[0; 4\pi/3]$.

7.25. At what values of the parameter a the equation

$$16^x - 3 \cdot 2^{3x+1} + 2 \cdot 4^{x+1} - (4 - 4a) \cdot 2^{x-1} - a^2 + 2a - 1 = 0$$

has exactly three different roots?

7.26. Find all the values of a such that the inequality

$$(x - \log_4 |a|) \cdot (x - \log_{|a|} 4) \cdot (x^2 - 10 \cdot 2^{-a}x + a^2 - 3) \geq 0$$

holds for any x.

7.27. Find all the values of a for which the equation

$$(\log_2(x+1) - \log_2(x-1))^2 - 2(\log_2(x+1) - \log_2(x-1))$$
$$- a^2 + 1 = 0$$

has exactly two different solutions.

7.28. Find all the values of a for which the equation

$$(|x-2| + |x+a|)^2 - 7(|x-2| + |x+a|) - 4a \cdot (4a - 7) = 0$$

has exactly two different solutions.

7.29. Find all the values of a for which the equation

$$\log_8^2 \left(\frac{x+a}{x-a}\right) - 12 \log_8 \left(\frac{x+a}{x-a}\right) + 35a^2 - 6a - 9 = 0$$

has exactly two different solutions.

7.30. Find all the values of a for which the equation

$$(x^2 + 2(a-2)x + a^2 - 4a)^2 + (a+5)(x^2 + 2(a-2)x + a^2 - 4a)$$
$$- a^2 + 8a + 2 = 0$$

has: (a) the only solution; (b) exactly two different solutions.

Tasks for Section 7.2

7.31. Find all the values of a for which the equation

$$2\cos 2x - 4a \cos x + a^2 + 2 = 0$$

has no solutions.

7.32. Find all the values of a for which the equation

$$a^2x^2 + 2a(\sqrt{2} - 1)x + \sqrt{x - 2} = 2\sqrt{2} - 3$$

has a solution.

7.33. Solve the system of equations

$$\begin{cases} \sqrt{x^2 + 3x + 2} - |y + 2| = 0, \\ 2\sqrt{y^2 + 4y + 4} + \sqrt{x^2 - x - 2} = 0. \end{cases}$$

7.34. Solve the system

$$\begin{cases} 2^{-x}y^4 - 2y^2 + 2^x \leq 0, \\ 8^x - y^4 + 2^x - 1 = 0. \end{cases}$$

7.35. Specify all the values of the parameter a for each of which the system

$$\begin{cases} x^2 + 2y^2 + 2y(x + a) + a^2 = 0, \\ 3^{-3-x} \cdot \log_3 y < 1 \end{cases}$$

has solutions, and find these solutions.

7.36. The number α is chosen so that the equation

$$\sqrt{x - \sqrt{3}} + \alpha^2 x^2 + 2\alpha x(\sqrt{6} - \sqrt{3}) = 6\sqrt{2} - 9$$

has a solution. Find this solution.

7.37. Find all triples of integers $(x; y; z)$ for which the following relation holds:

$$5x^2 + y^2 + 3z^2 - 2yz = 30.$$

7.38. Find all the values of a for which the equation

$$(x^2 - 6|x| - a)^2 + 12(x^2 - 6|x| - a) + 37 = \cos(18\pi/a)$$

has exactly two different solutions.

7.39. For what values of a does the equation

$$\left| \frac{x^2 - 4ax + 4a^2 - 1}{x - 2a} \right| + x^2 - 2x + 1 = 0$$

have at least one solution?

7.40. Solve the equation

$$(x-1)^6(\sin 4x + \sin 4)^{1/6} + (x+1)^6(\sin 2 - \sin 2x)^{1/6} = 0.$$

7.41. For each value of b, find all pairs of numbers (x, y) satisfying the equation

$$b\sin 2y + \log_4(x\sqrt[8]{1-4x^8}) = b^2.$$

7.42. For each value of a, find all pairs of numbers (x, y) satisfying the equation

$$a\cos 2x + \log_2(y\sqrt[12]{1-2y^{12}}) = a^2.$$

7.43. Find all pairs of real numbers a and b for which the equation

$$(3x - a^2 + ab - b^2)^2 + (2x^2 - a^2 - ab)^2 + x^2 + 9 = 6x$$

has at least one solution x.

7.44. Find all rational solutions to the equation

$$\sqrt{y \cdot (x+1)^2 - x^2 + x + 1} + \log^2_{(|y+2|/21)}\cos^2 \pi y = 0.$$

7.45. For what values of a the equation

$$\left(\sqrt{x^2 - 3ax + 8} + \sqrt{x^2 - 3ax + 6}\right)^x$$

$$+ \left(\sqrt{x^2 - 3ax + 8} - \sqrt{x^2 - 3ax + 6}\right)^x = 2(\sqrt{2})^x$$

has a unique solution?

7.46. Find the smallest of the values x for which there are numbers y and z satisfying the equation

$$x^2 + 2y^2 + z^2 + xy - xz - yz = 1.$$

7.47. Find all the solutions to the system

$$\begin{cases} \cos 10x - 2\sin 5x \geq 3 \cdot 4^t - 3 \cdot 2^{t+2} + 27/2, \\ \sqrt{(2-\sqrt{3})^{4t} + (2+\sqrt{3})^{4t} + 2 + 14\log_2(\cos 10x) + 6\cos 5x} \\ \geq (2t+1)^{1.5}. \end{cases}$$

7.48. Find the sum of the first hundred positive roots of the equation

$$\cos(8\pi x) + 2\cos(4\pi x) - \cos(2\pi x) - 2\sin(\pi x) + 3 = 0.$$

7.49. For each value of a, solve the equation

$$4 - \sin^2 x + \cos 4x + \cos 2x + 2 \sin 3x \sin 7x - \cos^2 7x - \cos^2 \pi a = 0.$$

7.50. Find all the values of a for each of which the system of inequalities

$$\begin{cases} 4^x - 2^{x+y} \le \dfrac{108a - 161}{2a - 3}, \\ 5 \cdot 2^{x+y} - 9 \cdot 4^y \ge 54 \end{cases}$$

has a solution.

7.51. For what values of a does the equation

$$(x^2 - x + a^2 + 2)^2 = 4a^2(2x^2 - x + 2)$$

have exactly three different solutions.

7.52. For each a, solve the equation

$$\cos^2 x + \cos^2 2x + \cos^2 3x + 2a(\cos x - \cos 2x + \cos 3x)$$
$$+ \cos 2x + \cos 4x + 2a^2 = 0.$$

Hints and Answers

7.1. $a = 12$.

7.2. $a \in (-2 - \sqrt{11}; -2 + \sqrt{11})$.

7.3. $a = \sqrt{3}/2$.

7.4. $a \in [-(7 + \sqrt{45})/2; -4 + 2\sqrt{3})$.

7.5. $a = -13, -4, -17/9$.

7.6. $a \in (1; +\infty)$.

7.7. $a \in (-1; 0] \cup \{(2 + 2\sqrt{13})/3\}$.

7.8. An empty set.

7.9. $x \in (-\infty; -2] \cup \{0\} \cup [7 + \sqrt{7}; +\infty)$.

7.10. $a = 2$.

7.11. $a \in (-\infty; 1] \cup \{5/4\} \cup [4/3; +\infty)$.

7.12. $|a| \le 3$.

7.13. $a \in (-\infty; 20]$.

7.14. $a \in \left(-\infty; \frac{-1 - \sqrt{43}}{3}\right] \cup [7; +\infty)$.

7.15. Addition to the set $\{a \in \mathbb{R} : a \in [-3; -1) \cup \{0; 4; \frac{-7 \pm \sqrt{33}}{2}\}\}$.

7.16. $a \in (0; 1/8)$.

7.17. $a > -2\sqrt{2}$.

7.18. $a = 2$. **Hint:** A quadratic equation must have at least one non-negative root.

7.19. (1) $a \in (1; (2 + \sqrt{13})/4]$; (2) $b = 7/3$.

7.20. $a \in (2; +\infty)$.

7.21. $p \in (-\infty; -5/2]$.

7.22. If $a < 0$, there are no solutions; if $a = 0$, then $x \in (0; +\infty)$; if $a > 0$, then $x \in [-a/3; 0) \cup (8a; +\infty)$.

7.23. $b \in (-\infty; -1/(2\sqrt{2})] \cup [-1/4; 1/4] \cup [1/(2\sqrt{2}); +\infty)$.

7.24. $a \in (-1/3; 3/10] \cup \{1\}$.

7.25. $a \in (0; 1) \cup (1; 4) \cup (4; 5)$. **Hint:** Factorize and explore the two quadratic equations.

7.26. $a = 4$, $a = 2$.

7.27. $a \in (-1; 0) \cup (0; 1)$.

7.28. $a \in (-\infty; 2/3) \cup \{7/8\} \cup (1; +\infty)$.

7.29. $a \in ((3 - 12\sqrt{11})/35; -3/7) \cup (-3/7; 0) \cup (0; 3/5) \cup (3/5; (3 + 12\sqrt{11})/35)$.

7.30. (a) $a = 2 + \sqrt{2}$, (b) $a \in (-\infty; 2 - \sqrt{2}) \cup \{1\} \cup (2 + \sqrt{2}; +\infty)$.

7.31. $a \in (-\infty; -2) \cup (2; +\infty)$.

7.32. $a = (1 - \sqrt{2})/2$, $x = 2$.

7.33. $x = -1$, $y = -2$.

7.34. $x = 0$, $y = \pm 1$.

7.35. $x = a$, $y = -a$ if $a \in (-3; 0)$.

7.36. $x = \sqrt{3}$.

7.37. $(x; y; z) = (1; 5; 0)$; $(1; -5; 0)$; $(-1; 5; 0)$; $(-1; -5; 0)$.

7.38. $a = -3$, $a = 9$.

7.39. $a = 0$, $a = 1$.

7.40. $x = -1$; $-1 + \pi/2 + \pi n$, $n \in \mathbb{Z}$.

7.41. If $b = -1/2$, the solutions are $(1/\sqrt[8]{8}; -\pi/4 + \pi k)$, $k \in \mathbb{Z}$; if $b = 1/2$, then $(1/\sqrt[8]{8}; \pi/4 + \pi m)$, $m \in \mathbb{Z}$; if $b \neq \pm 1/2$, there are no solutions.

7.42. If $a = -1/2$ then the solutions are $(\pi/2 + \pi k; 1/\sqrt[6]{2})$, $k \in \mathbb{Z}$; if $a = 1/2$, then $(\pi m; 1/\sqrt[6]{2})$, $m \in \mathbb{Z}$; there are no solutions for $a \neq \pm 1/2$.

7.43. $(3; 3)$; $(-3; -3)$; $(2\sqrt{3}; \sqrt{3})$; $(-2\sqrt{3}; -\sqrt{3})$.

7.44. $(x; y) = (-2/3; 1)$; $(-1 - 1/(l-1); l^2 + l - 1)$; $(-1 + 1/(l+2); l^2 + l - 1)$, $l \in \mathbb{Z}$, $l \neq -5, -2, 1, 4$.

7.45. $a \in (-2\sqrt{6}/3; 2\sqrt{6}/3)$.

7.46. $x = -\sqrt{7/5}$. **Hint:** Select the full square first using the variable z.

7.47. $t = 1$, $x = -\pi/30 + 2\pi n/5$, $n \in \mathbb{Z}$.

7.48. 4950.

7.49. If $a \in \mathbb{Z}$, then $x = \pi/4 + \pi k/2$, $k \in \mathbb{Z}$; there are no solutions for other a.

7.50. $a \in (3/2; +\infty)$. **Hint:** Put $u = 2^x$, $v = 2^y$. Subtract the second inequality from the first, and prove that with the resulting constraint on a, there is always a solution.

7.51. $a = \pm\sqrt{2}, \pm(\sqrt{15}+1)/4$.

7.52. If $a = 0$, then $x = \pi/4 + \pi m/2$; if $a = -1/2$, then $x = \pm 2\pi/3 + 2\pi n$, where $m, n \in \mathbb{Z}$; there are no solutions for $a \neq 0, -1/2$.

Chapter 8

Solving Problems Using a Geometric Image

8.1 Graphs of Some Elementary Functions and Solving Problems with Their Help

Let us recall some equations of curves and graphs of functions.

I(a). Let us start with the general equation of the line

$$ax + by + c = 0, \quad a^2 + b^2 \neq 0,$$

which is called its canonical equation (see Fig. 8.1).

I(b) The equation of a straight line passing through two different points, (x_1, y_1) and (x_2, y_2), is written as

$$\frac{x - x_1}{x_2 - x_1} = \frac{y - y_1}{y_2 - y_1}, \quad x_1 \neq x_2, y_1 \neq y_2,$$

(see Fig. 8.3). In the case of $x_2 = x_1$, the equation of the line takes the form $x = x_1$, and in the case of $y_2 = y_1$, it takes the form $y = y_1$.

I(c) The graph of the function $y = |x - a|$ is shown in Fig. 8.2. (In general, to plot the function $y = |f(x)|$ according to a given graph of the function $y = f(x)$, all values of the function $y = f(x)$ should be replaced with their absolute values, for which it is necessary to replace the negative values of the function $f(x)$ by $-f(x)$, i.e., reflect the points of the graph with a negative ordinate symmetrically with respect to the straight line $y = 0$; see Figs. 8.4 and 8.5.)

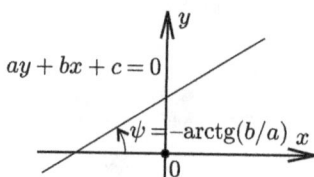

Fig. 8.1.   Straight line.

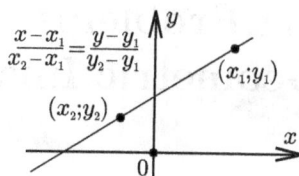

Fig. 8.2.   Graph of straight line with two points.

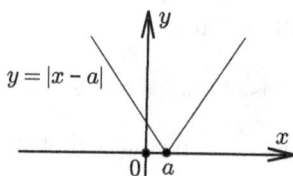

Fig. 8.3.   Graph of modulus.

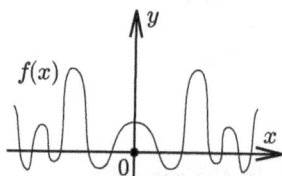

Fig. 8.4.   Graph $f(x)$.

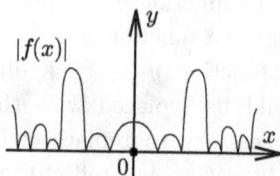

Fig. 8.5.   Graph $|f(x)|$.

II. The parabola equation has the form (see Section 7.1)

$$y = ax^2 + bx + c, \quad a \neq 0.$$

III. The equation of a circle centered at $(x_0; y_0)$ and with radius R is (see Fig. 8.6)

$$(x - x_0)^2 + (y - y_0)^2 = R^2.$$

IV. The distance between two points see Fig. 8.7 $(x_1; y_1)$ and $(x_2; y_2)$ on the plane is

$$d = \sqrt{(x_1 - x_2)^2 + (y_1 - y_2)^2}.$$

V. The hyperbola equation in the simplest case has the form $y = \frac{1}{x}$ (see Fig. 8.8). Its vertical asymptote is $x = 0$, and its horizontal asymptote

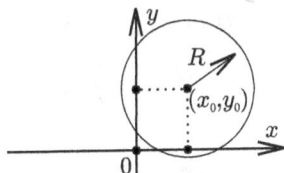

Fig. 8.6. $(x - x_0)^2 + (y - y_0)^2 = R^2.$

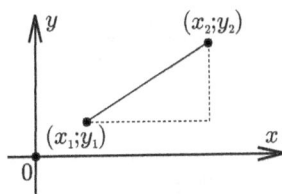

Fig. 8.7.   Distance between points.

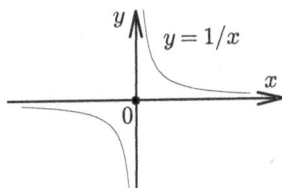

Fig. 8.8.   Graph $f(x) = 1/x.$

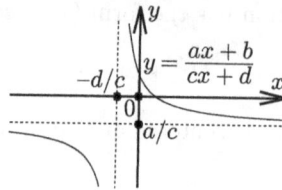

Fig. 8.9.   Graph $f(x) = \frac{ax+b}{cx+d}$.

is $y = 0$. Similarly, you can plot an arbitrary fractional linear function $y = \frac{ax+b}{cx+d}$ (again the graph will be a hyperbola), $a, b, c, d \in \mathbb{R}$, $c \neq 0$. Indeed, from the equality

$$\frac{ax+b}{cx+d} = \frac{\frac{a}{c}(cx+d)+b-\frac{ad}{c}}{cx+d} = \frac{a}{c} + \frac{bc-ad}{c(cx+d)},$$

we conclude that the graph of a fractional linear function can be obtained from the hyperbola $y = 1/x$ by shifting and stretching (see Fig. 8.9).

Example 8.1. Find all values of a for which the equation

$$4x - |3x - |x + a|| = 9|x - 1|$$

has at least one root.

Proof. Consider the function

$$f(x) = 9|x - 1| - 4x + |3x - |x + a||.$$

By expanding the moduli, we get a finite number of intervals, on each of which it is some linear function. The coefficient at the first modulus exceeds in absolute value the sum of the remaining coefficients at x, with whatever sign we do not disclose the remaining moduli. Indeed, $9 - 4 - 3 - 1 = 1 > 0$. Therefore, on all intervals lying to the left of the point $x = 1$, the coefficient at x is negative, and on all intervals to the right of the point $x = 1$, the coefficient at x is positive. This means that the function $f(x)$ decreases for $x < 1$ and increases for $x > 1$, and $x = 1$ is the minimum point (see Fig. 8.10). Therefore, in order for the equation $f(x) = 0$ to have at least one

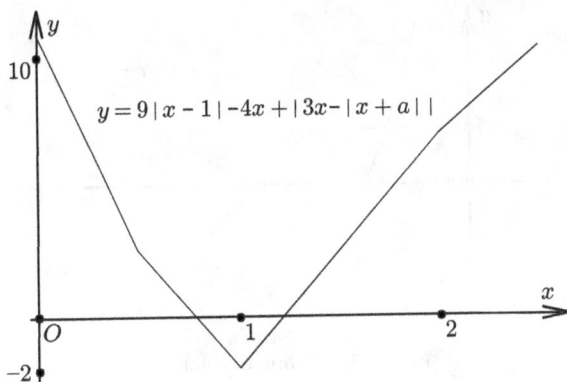

Fig. 8.10.   Case: $a = -2$.

root, it is necessary and sufficient that the following condition is fulfilled: $\min f(x) \le 0$, i.e., $f(1) \le 0$. We introduce the notation $t = |1 + a|$. Then,

$$f(1) \le 0 \Longleftrightarrow |3 - |1 + a|| - 4 \le 0 \Longleftrightarrow |3 - t| \le 4$$
$$\Longleftrightarrow (3 - t)^2 - 4^2 \le 0 \Longleftrightarrow (-1 - t)(7 - t) \le 0$$
$$\Longleftrightarrow (1 + t)(t - 7) \le 0 \Longleftrightarrow t \in [-1; 7].$$

Now, for a, we get the inequality $|1 + a| \le 7$, solving which we arrive at the answer: $a \in [-8; 6]$.

Answer: $a \in [-8; 6]$. □

Example 8.2. Find all the values of a for which the equation

$$(a + 6x - x^2 - 8)(a - 1 + |x - 3|) = 0$$

has exactly three different solutions.

Proof. Let us draw (see Fig. 8.11) on the plane $(x; a)$ a parabola given by the equation $a + 6x - x^2 - 8 = 0$ (equivalent to the equation $a = (x-3)^2 - 1$) and a polyline $a = 1 - |x - 3|$. The condition of the problem is satisfied by the values $a = \pm 1$ and only them.

Answer: $a = \pm 1$. □

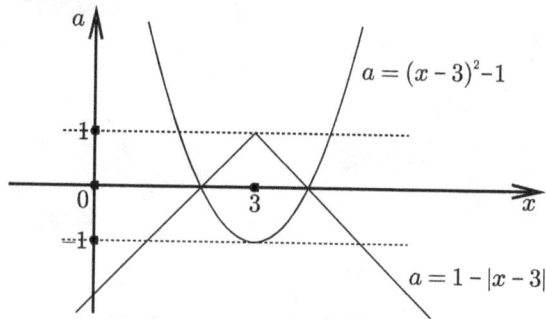

Fig. 8.11.   Graph of solution.

Example 8.3. Find all the values of a for which the equation

$$\left|(2x-a)^2 - |x| - 28\right| + 2|x| = 16$$

has three different solutions.

Proof. Let us make equivalent transformations of the original equation:

$$\left|(2x-a)^2 - |x| - 28\right| = 16 - 2|x|$$

$$\Longleftrightarrow \begin{cases} 16 - 2|x| \geq 0, \\ \left[\begin{array}{l} (2x-a)^2 - |x| - 28 = 16 - 2|x|, \\ (2x-a)^2 - |x| - 28 = -16 + 2|x|. \end{array}\right. \end{cases}$$

$$\Longleftrightarrow \begin{cases} x \in [-8; 8], \\ \left[\begin{array}{l} (2x-a)^2 = 44 - |x|, \\ (2x-a)^2 = 3|x| + 12. \end{array}\right. \end{cases}$$

Let's draw the graphs of the functions $y = 44 - |x|$ and $y = 3|x| + 12$, for $|x| \leq 8$, as shown in Fig. 8.12 and find those parabolas of the form $y = (2x-a)^2$ that satisfy the condition of the problem. There will be three solutions in the case when the parabola passes through the intersection point of the graphs of the functions $y = 44 - |x|$ and $y = 3|x| + 12$, and the vertex of the parabola will belong to the segment $[-8; 8]$.

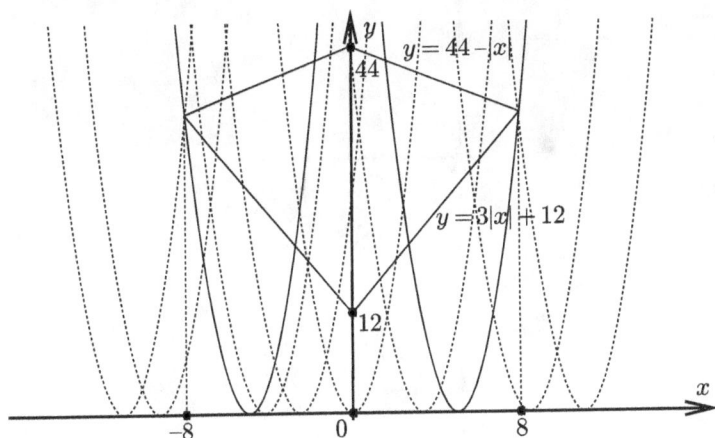

Fig. 8.12. Graphs.

Since the functions $y = 44 - |x|$ and $y = 3|x| + 12$ intersect at the points $(\pm 8; 36)$, the desired parabolas satisfy the system

$$\begin{cases} (2 \cdot (\pm 8) - a)^2 = 36, \\ a/2 \in [-8; 8]. \end{cases} \Longleftrightarrow \begin{cases} a = \pm 22; \pm 10, \\ a/2 \in [-8; 8]. \end{cases} \Longleftrightarrow a = \pm 10.$$

Answer: $a = \pm 10$. $\qquad\qquad\qquad\qquad\qquad\qquad\qquad\qquad\qquad\qquad$ $\square$

Example 8.4. Find all the values of a for which the function

$$f(x) = x^2 - 2|x - a^2| - 8x$$

has at least one maximum point.

Proof. 1. The function $f(x)$ has the following form:

  i. if $x \geq a^2$, then $f(x) = x^2 - 10x + 2a^2 = (x - 5)^2 - 2a^2 - 25$, so the graph of the function $f(x)$ is part of a parabola with branches pointing upward and the axis symmetry $x = 5$;

  ii. if $x \leq a^2$, then $f(x) = x^2 - 6x - 2a^2 = (x - 3)^2 + (2a^2 - 9)$, so the graph of the function $f(x)$ is part of a parabola with branches pointing upward, and the axis of symmetry $x = 3$.

All possible types of graphs of the function $f(x)$ for different values of a^2 are shown in Figs. 8.13 and 8.14.

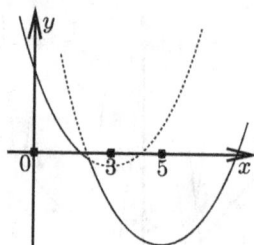

Fig. 8.13. $a^2 \leq 3$.

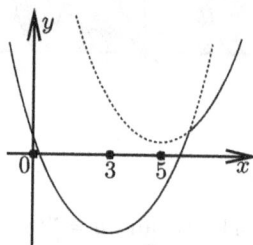

Fig. 8.14. $a^2 \geq 5$.

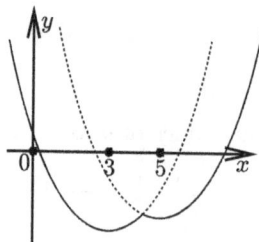

Fig. 8.15. $3 < a^2 < 5$.

2. None of the functions shown in Figs. 8.13 and 8.14 have maximum points. Indeed, the graphs of both functions pass through the point $(a^2; f(a^2))$, and the first of them decreases in the vicinity of this point, whereas the second function increases.

3. Thus, the only maximum point of the function $f(x)$ is the point $x = a^2$ (see Fig. 8.15), and if and only if $3 < a^2 < 5 \iff \sqrt{3} < |a| < \sqrt{5}$.

Answer: $a \in (-\sqrt{5}; -\sqrt{3}) \cup (\sqrt{3}; \sqrt{5})$. $\qquad\qquad\qquad\square$

Example 8.5. Find all the values of c for which the system

$$\begin{cases} y = ||x + 3| - 1|, \\ x^2 + y^2 = 2cy - c^2 - 4x - \dfrac{7}{2} \end{cases}$$

has exactly two different solutions.

Proof. Let us transform the system to the form

$$\begin{cases} y = ||x + 3| - 1|, \\ (x + 2)^2 + (y - c)^2 = \dfrac{1}{2}. \end{cases}$$

Let us describe the plotting of the function $y = ||x + 3| - 1|$. We plot the function $y = |x + 3|$ (see Fig. 8.16), then we shift the graph one unit down (see Fig. 8.17). To plot the graph $y = ||x + 3| - 1|$, the part of the graph of the function $y = |x + 3| - 1$ lying below the Ox axis should be displayed symmetrically relative to the Ox axis (see Fig. 8.18). The second equation in the system defines a circle centered at $(-2; c)$ and with a radius of $1/\sqrt{2}$. Figure 8.19 shows the possible locations of the curves from the example. We note that the cases are suitable for us when the circle touches

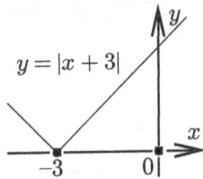

Fig. 8.16.   Graph of $y = |x + 3|$.

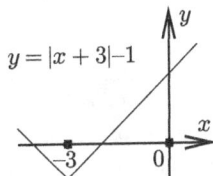

Fig. 8.17.   Graph of $y = |x + 3| - 1$.

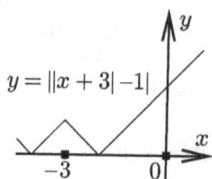

Fig. 8.18.   Graph of $y = ||x + 3| - 1|$.

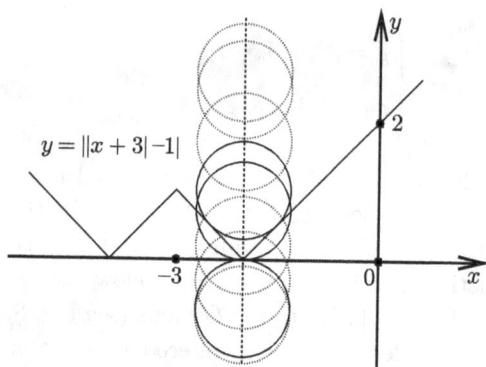

Fig. 8.19.   Solution depending on the parameter.

the graph $y = ||x + 3| - 1|$ at $c = 1$ and crosses the graph at two points: $c \in (-1/\sqrt{(2)}; 1/\sqrt{(2)})$.

Answer: $a \in (-1/\sqrt{2}; 1/\sqrt{2}) \cup \{1\}$.   □

Example 8.6. Find all the values of a for which the set of solutions to the inequality

$$\sqrt{5 - x} \leq 3 - |x - a|$$

is a segment.

Proof. Let us graphically represent the solutions of the inequality $\sqrt{5 - x} \leq 3 - |x - a|$, i.e., find those points for which the graph of the function $y = 3 - |x - a|$ ("corner") is located above the graph of the function $y = \sqrt{5 - x}$ (or they intersect). Figure 8.20 shows possible cases of mutual arrangement of these graphs. If the abscissa of the vertex of the "corner" is

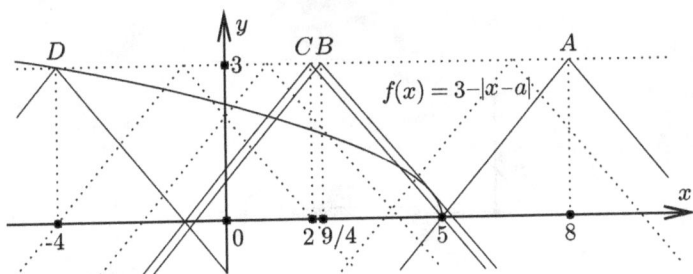

Fig. 8.20. Possible cases of mutual arrangement of graphs.

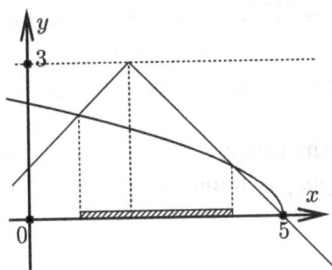

Fig. 8.21. Case $a = 2$.

located to the left of the abscissa of the point D, then the original inequality has no solutions. If the vertex of the "corner" coincides with the point D, then the original inequality has a unique solution for which $\sqrt{5 - x} = 3$, i.e., $x = -4$, and since $|-4 - a| = 0$, we get $a = -4$.

When moving the vertex of the "corner" to the right, if the abscissa of the vertex is greater than the abscissa of the point D ($x = -4$) but less than the abscissa of the point C ($x = 2$), the set of solutions to the inequality is a segment. Because the abscissa of the vertex of the "corner" coincides with the value a, we get that for $-4 < a < 2$, the condition of the problem is fulfilled.

At $a = 2$, the initial inequality has a set of solutions consisting of a segment and a separately located point $x = 5$, so the condition of the problem is not fulfilled (see Fig. 8.21).

Figure 8.22 shows in more detail the situation when the abscissa of the vertex of the "corner" is located between the points C and B (the point B corresponds to the case when the right part of the "corner" touches the

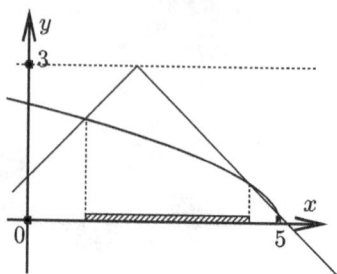

Fig. 8.22. Case $a \in (2; \frac{9}{4})$.

graph of the function $y = \sqrt{5-x}$; the corresponding value of a is found in the following). In this case, the set of solutions to the inequality consists of two segments. To find the abscissa of the vertex B, we propose two methods.

1. In the first method, the tangent point of two graphs of the differentiable functions $f(x)$ and $g(x)$ is found from the conditions $f(x) = g(x)$ and $f'(x) = g'(x)$:

$$\begin{cases} 3 - (x - a) = \sqrt{5-x}, \\ -1 = -\dfrac{1}{2\sqrt{5-x}}. \end{cases} \Longleftrightarrow \begin{cases} x = 19/4, \\ a = 9/4. \end{cases}$$

2. Another way to find the tangent point is to find a from the condition of uniqueness of the solution of the equation $\pm\sqrt{5-x} = 3 - (x - a)$ (i.e., the intersection of a straight line and a parabola; here, by putting a sign, we restored the whole parabola):

$$\pm\sqrt{5-x} = 3 + a - x \Longleftrightarrow \begin{cases} x^2 - (5 + 2a)x + a^2 + 6a + 4 = 0, \\ 5 - x \geq 0. \end{cases}$$

From the condition for the discriminant of the quadratic equation $D = -4a + 9 = 0$, we obtain the solution $a = 9/4$, $x = 19/4$.

If the abscissa of the vertex of the "corner" is located between the abscissa points A and B, then the set of solutions is again a segment. Starting from the point A, the set of solutions either consists of a single point or is empty.

The abscissa of the point A is equal to 8. Therefore, the set of solutions of the original inequality is a segment and for $a \in [9/4; 8)$. Combining the parts of the answer, we get $a \in (-4; 2) \cup [9/4; 8)$.

Answer: $a \in (-4; 2) \cup [9/4; 8)$. $\qquad\square$

8.2 More Complex Graphs

In addition to what was said in the previous paragraph, we add the following:

VI. The angle α between the curves $y = f(x)$ and $y = g(x)$ (see Fig. 8.23) at the point of their intersection x_0 is the angle between their tangents at the point x_0. In particular, if the angle between the curves $y = f(x)$ and $y = g(x)$ at their intersection point (x_0, y_0) is zero, then the tangents at the point (x_0, y_0) for curves $y = f(x)$ and $y = g(x)$ coincide (see Fig. 8.24), and we say that the curves $y = f(x)$ and $y = g(x)$ are tangent to each other at the point (x_0, y_0). The condition of tangency of

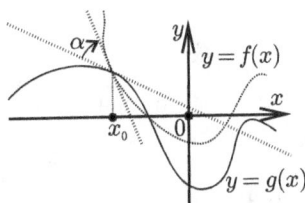

Fig. 8.23.   Angle between the curves.

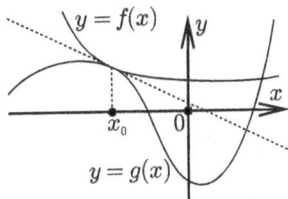

Fig. 8.24.   The common tangent.

two differentiable curves at the point with the abscissa x_0 is equivalent to the following system of equations:

$$\begin{cases} f(x_0) = g(x_0), \\ f'(x_0) = g'(x_0). \end{cases}$$

Example 8.7. Find all the values of a for which the system of equations

$$\begin{cases} (xy + x + y)(y + x^2) = 0, \\ y = ax - 1 \end{cases}$$

has (a) exactly two different solutions and (b) exactly four different solutions.

Proof. The first equation is satisfied by the hyperbola points $xy + x + y = 0$ (or $(x+1)(y+1) = 1$, from where $y = -x/(x+1)$) and parabola $y = -x^2$ (see Fig. 8.25). Find the intersection points of the hyperbola with the parabola:

$$-\frac{x}{x+1} = -x^2 \iff \frac{x}{x+1} \cdot (x^2 + x - 1) = 0 \iff x_1 = 0 \ x_{2,3}$$
$$= (-1 \pm \sqrt{5})/2.$$

The coordinates of the intersection points of the hyperbola with the parabola are denoted by $(x_{1,2,3}; y_{1,2,3})$, and the corresponding values of $y_{1,2,3}$ are found from the equation $y = -x^2$. The roots of $x_{2,3}$ satisfy the equation $x_{2,3}^2 + x_{2,3} - 1 = 0$, from where $y_{2,3} = -x_{2,3}^2 = x_{2,3} - 1$. Thus, the corresponding intersection points of a hyperbola with a parabola lie on a straight line A, given by the equation $y = x - 1$ (see Fig. 8.25). For different values of a, the graphs of the functions given by the equation $y = ax - 1$ are straight lines passing through the point $(0; -1)$. Denote by B the line $x = 0$ and by D, denote the line $y = -1$.

The line $y = ax - 1$ intersects with the parabola $y = -x^2$ at two points for any a since the equation $-x^2 = ax - 1$ has a positive discriminant. Let's find the values of a for which a line of the form $y = ax - 1$ touches the hyperbola. To do this, it is enough to find those a at which the intersection point of the line with the hyperbola is unique, i.e., the equation $-\frac{x}{x+1} = ax - 1$ has a unique solution. Since this equation reduces to a square one, calculating the discriminant and equating it to zero, we find $a = -4$ and the tangent point $(x; y) = (-1/2; 1)$. The point of contact could also be

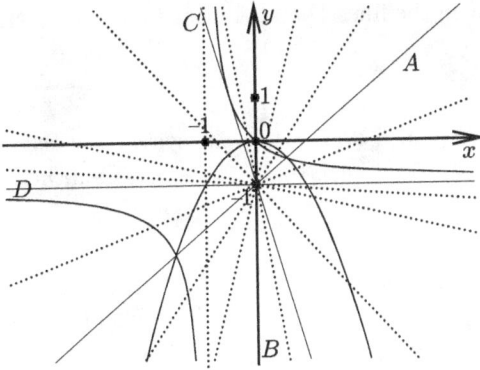

Fig. 8.25. The general solution.

found from the system, which means that both the values of functions and the values of their derivatives are equal:

$$\begin{cases} ax - 1 = -1 + \dfrac{1}{x+1}, \\ a = -\dfrac{1}{(x+1)^2}. \end{cases}$$

Thus, for different values of a we get the following (see Fig. 8.25):

1. The straight line A (see Fig. 8.25) has two intersection points with curves that are solutions of the first equation in the system (which corresponds to $a = 1$).
2. The line B (see Fig. 8.25) has one intersection point with curves that are solutions of the first equation (the line B does not belong to the family of lines $y = ax - 1$ under any a, although it passes through the point $(0; -1)$).
3. The line between the lines A and B gives four intersection points (which corresponds to $a \in (1; +\infty)$).
4. The line C (see Fig. 8.25) gives three intersection points (which corresponds to $a = -4$).
5. The line between the lines B and C gives four intersection points (which corresponds to $a \in (-\infty; -4)$).
6. The line D (see Fig. 8.25) gives two intersection points (which corresponds to $a = 0$).
7. The line between the lines C and D gives two intersection points (which corresponds to $a \in (0; -4)$).

8. The line between the lines D and A gives four intersection points (which corresponds to $a \in (0; 1)$).

Answer: (a) If $a \in (-4; 0] \cup \{1\}$, then the system has exactly two different solutions; (b) if $a \in (-\infty; -4) \cup (0; 1) \cup (1; +\infty)$, then there are exactly four different solutions. □

Example 8.8. Find all the values of a for which the system

$$\begin{cases} y^2 + a = x, \\ |x| + |y| + |x - y| = 2 \end{cases}$$

has (a) exactly one solution and (b) exactly four different solutions.

Proof. Let us construct a polyline given by the equation $|x| + |y| + |x - y| = 2$. Depending on which signs the values x, y, and $x - y$ have, consider six areas (see Fig. 8.26). In each of these six regions, the curve defined by the equation $|x| + |y| + |x - y| = 2$ is a segment. To find the ends of these segments, we use the equation $|x| + |y| + |x - y| = 2$, from which it follows that:

1. if $x = 0$, then $y = \pm 1$;
2. if $y = 0$, then $x = \pm 1$;
3. if $x = y$, then $x = y = 1$ or $x = y = -1$.

Thus, we found two points in each of the six regions (see Fig. 8.27). These points are the endpoints of the desired segments (see Fig. 8.28).

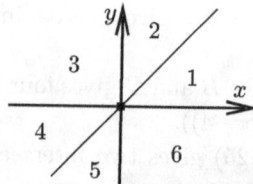

Fig. 8.26.   Six regions.

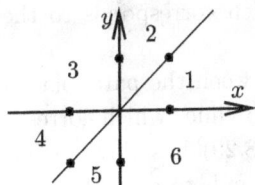

Fig. 8.27. Six points.

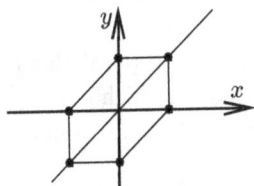

Fig. 8.28. Graph of $|x| + |y| + |x - y| = 2$.

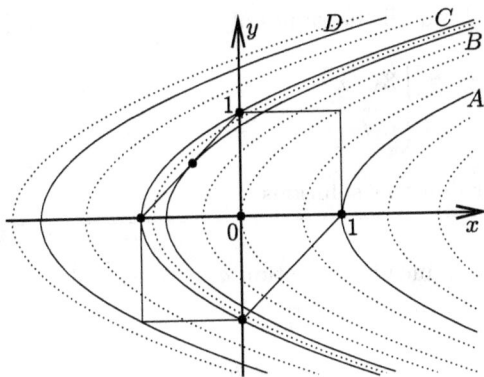

Fig. 8.29. The general solution.

Note that the first equation in the system $y^2 + a = x$ defines a parabola. The following cases are possible:

1. Parabola A (see Fig. 8.29) has one intersection point with the polyline given by the second equation in the system (which corresponds to $a = 1$).
2. Parabola B (see Fig. 8.29) has three points of intersection with the polyline (which corresponds to $a = a_0$). The value of a_0 is found from the condition that the line $y - x = 1$ is tangent to the parabola $x = y^2 + a$.

Thus, $a_0 = -3/4$, which corresponds to the contact point $(x_0; y_0) = (-1/2; 1/2)$.

3. A parabola located between the parabolas A and B has two points of intersection with the polyline (which corresponds to $a \in (-3/4; 1)$).

4. Parabola C (see Fig. 8.29) has three points of intersection with the polyline (which corresponds to $a = -1$).

5. A parabola located between the parabolas B and C has four points of intersection with the polyline (which corresponds to $a \in (-1; -3/4)$).

6. Parabola D (see Fig. 8.29) has one intersection point with the polyline (which corresponds to $a = -2$).

7. A parabola located between the parabolas C and D has two points of intersection with the polyline (which corresponds to $a \in (-2; -1)$).

8. In the case of $a > 1$ or $a < -2$, there are no intersections (see Fig. 8.29).

Answer: (a) $a = -2$, $a = 1$; (b) $a \in (-1; -3/4)$.   □

Example 8.9. Find all the values of c for which the system

$$\begin{cases} 3\sqrt{|x+4|} + \sqrt{|y-3|} = 1, \\ 81(x+4)^2 + y^2 + c = 6y \end{cases}$$

has exactly four different solutions.

Proof. Let us introduce the notation $a = 3\sqrt{|x+4|}$, $b = \sqrt{|y-3|}$. Then, the system can be written as

$$\begin{cases} a \geq 0, \quad b \geq 0, \\ a + b = 1, \\ a^4 + b^4 = 9 - c. \end{cases} \tag{8.1}$$

This system can also be solved using symmetry considerations, but in this section we analyze the graphical interpretation of this example (see Fig. 8.30). Consider the following possible cases:

1. Let $(a_0; 0)$, $a_0 > 0$ be the solution of system (8.1). In this case, the original system has two solutions: $(x; y) = (-4 \pm (a_0/3)^2; 3)$.

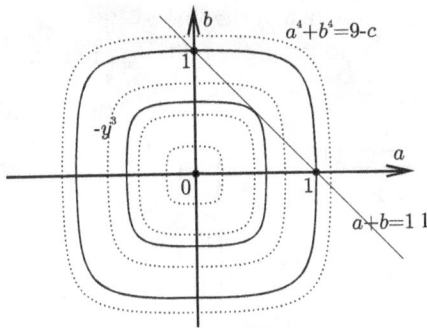

Fig. 8.30. Graphs of curves from (8.1).

2. Let $(0; b_0)$, $b_0 > 0$ be the solution of system (8.1). In this case, the original system has two solutions: $(x; y) = (-4; 3 \pm (b_0)^2)$.
3. Let $(a_0; b_0)$, $a_0 > 0$, $b_0 > 0$ be the solution of system (8.1). In this case, the original system has four solutions: $(x; y) = (-4 \pm (a_0/3)^2; 3 \pm b_0^2)$.
4. If $(a_0; b_0)$, $a_0 \le 0$, $b_0 \le 0$, then $(a_0; b_0)$ does not satisfy system (8.1). There are no solutions.

It follows from the above that there can be four solutions only in the following two cases (see Fig. 8.30):

1. The line $a + b = 1$ intersects the graph of the function $a^4 + b^4 = 9 - c$ at points lying on the coordinate axes.
2. The line $a + b = 1$ touches the graph of the function $a^4 + b^4 = 9 - c$ at the point $(a_0; b_0)$, where $a_0 > 0$, $b_0 > 0$.

The first case is possible when the line $a + b = 1$ intersects the graph of the function $a^4 + b^4 = 9 - c$ at the points $(1; 0)$ and $(0; 1)$, which means $1 = 9 - c$, i.e., $c = 8$.

Let us analyze the second case, i.e., the case of contact. We express b from the system, taking into account that the value of b is non-negative:

$$\begin{cases} b = 1 - a, \\ b = \sqrt[4]{9 - c - a^4}. \end{cases}$$

The tangent case is possible when the line $b = 1 - a$ is tangent to the graph $b = \sqrt[4]{9 - c - a^4}$, i.e., the following conditions are fulfilled:

$$\begin{cases} 1 - a = \sqrt[4]{9 - c - a^4}, \\ (1 - a)' = (\sqrt[4]{9 - c - a^4})'. \end{cases} \iff \begin{cases} 1 - a = (9 - c - a^4)^{1/4}, \\ -1 = \dfrac{1}{4}(9 - c - a^4)^{-3/4}(-4a^3). \end{cases}$$

$$\iff \begin{cases} 1 - a = (9 - c - a^4)^{1/4}, \\ 1 = (1 - a)^{-3} \cdot a^3. \end{cases}$$

From the above, we get that $\frac{a}{1-a} = 1$ or $a = \frac{1}{2}$ and $c = -71/8$.

Answer: $c = 71/8,\ 8$. □

Example 8.10. Solve the inequality

$$\arcsin(\sin x) + 3\arccos(\cos x) \geq 3x - 18.$$

Proof. Note that the functions $\arcsin(\sin x)$ and $\arccos(\cos x)$ are periodic with a period of 2π (see Figs. 8.31 and 8.32). In particular, we have

$$\arcsin(\sin x) = \begin{cases} x - 2\pi k, & x \in [-\pi/2 + 2\pi k; \pi/2 + 2\pi k], \quad k \in \mathbb{Z}, \\ \pi - x + 2\pi k, & x \in [\pi/2 + 2\pi k; 3\pi/2 + 2\pi k], \quad k \in \mathbb{Z} \end{cases}$$

and

$$\arccos(\cos x) = \begin{cases} x - 2\pi k, & x \in [2\pi k; \pi + 2\pi k], \quad k \in \mathbb{Z}, \\ -x + 2\pi k, & x \in [-\pi + 2\pi k; 2\pi k], \quad k \in \mathbb{Z}. \end{cases}$$

Let's plot the function $f(x) = \arcsin(\sin x) + 3\arccos(\cos x)$ on the period, i.e., on the segment $[0; 2\pi]$. To do this, we plot the points $(x; f(x))$ with

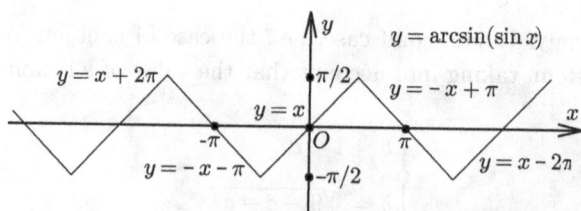

Fig. 8.31.   Graph of $y = \arcsin(\sin x)$.

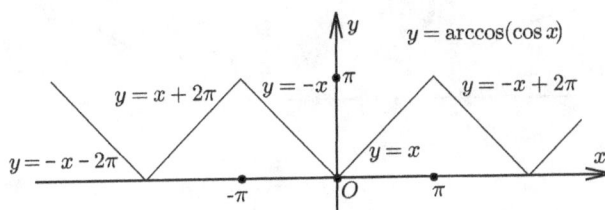

Fig. 8.32. Graph of $y = \arccos(\cos x)$.

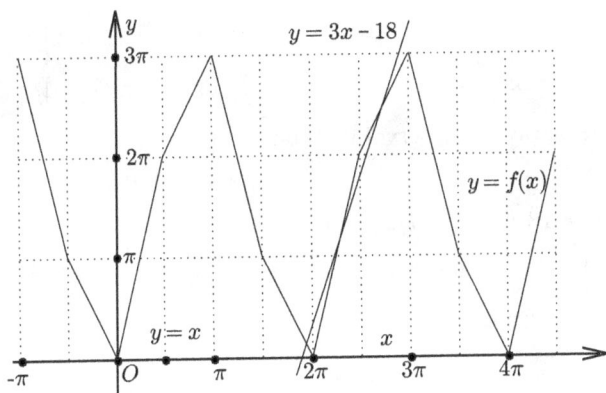

Fig. 8.33. Graphs of $y = f(x)$ and $y = 3x - 18$.

abscissae $x = 0$, $\pi/2$, π, $3\pi/2$, and 2π on the coordinate plane and connect them with segments. Let's continue the graph for the entire straight line using the fact that the original function is periodic with a period of 2π. Then, we plot a straight line $y = 3x - 18$ (see Fig. 8.33). Solve the equation:

$$\arcsin(\sin x) + 3\arccos(\cos x) = 3x - 18,$$

and then we solve the initial inequality using the interval method. Since the function $f(x)$ satisfies the conditions $0 \leq f(x) \leq 3\pi$, on the set $(-\infty; 3\pi/2) \cup (3\pi; +\infty)$, the equation has no solutions (the values of the function $g(x) = 3x - 18$ in these sections do not belong to the segment $[0; 3\pi]$). For the function $f(x)$, we have

$$f(x) = \arcsin(\sin x) + 3\arccos(\cos x) = \begin{cases} -2x + 4\pi, & x \in [3\pi/2; 2\pi], \\ 4x - 8\pi, & x \in [2\pi; 5\pi/2], \\ 2x - 3\pi, & x \in [5\pi/2; 3\pi]. \end{cases}$$

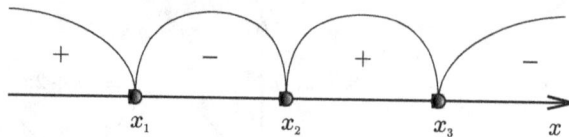

Fig. 8.34. $f(x) - g(x) \geq 0$.

Hence,

$$-2x + 4\pi = 3x - 18 \quad \Longleftrightarrow \quad x_1 = (4\pi + 18)/5 \in [3\pi/2; 2\pi],$$

$$4x - 8\pi = 3x - 18 \quad \Longleftrightarrow \quad x_2 = 8\pi - 18 \in [2\pi; 5\pi/2],$$

$$2x - 3\pi = 3x - 18 \quad \Longleftrightarrow \quad x_3 = 18 - 3\pi \in [5\pi/2; 3\pi].$$

It remains to apply the interval method to the inequality $f(x) - g(x) \geq 0$ (see Fig. 8.34):

$$f(0) - g(0) = 18 \quad \Longrightarrow \quad f(x) - g(x) \geq 0, \quad x \in (-\infty; x_1],$$

$$f(2\pi) - g(2\pi) = 18 - 6\pi < 0 \quad \Longrightarrow \quad f(x) - g(x) < 0, \quad x \in (x_1; x_2),$$

$$f(5\pi/2) - g(5\pi/2) = 2\pi - 15\pi/2 + 18 = 18 - 11\pi/2 > 0$$

$$\Longrightarrow \quad f(x) - g(x) \geq 0, \quad x \in [x_2; x_3],$$

$$f(4\pi) - g(4\pi) = 18 - 12\pi < 0 \quad \Longrightarrow \quad f(x) - g(x) < 0, \quad x \in (x_3; +\infty).$$

So, we get the answer.

Answer: $(-\infty; (4\pi + 18)/5] \cup [8\pi - 18; 18 - 3\pi]$. $\qquad \square$

Example 8.11. Find all positive a for which the equation

$$\frac{2\pi a + \arcsin(\sin x) + 2\arccos(\cos x) - ax}{\tan^2 x + 1} = 0$$

has exactly three different solutions belonging to the set $(-\infty; 7\pi]$.

Proof. The domain of definition is $x \in \mathbb{R} \backslash \{\pi/2 + \pi n\}_{n=1}^{+\infty}$. On the domain of definition, we solve the equation

$$\arcsin(\sin x) + 2\arccos(\cos x) = ax - 2\pi a.$$

The function

$$f(x) = \arcsin(\sin x) + 2\arccos(\cos x).$$

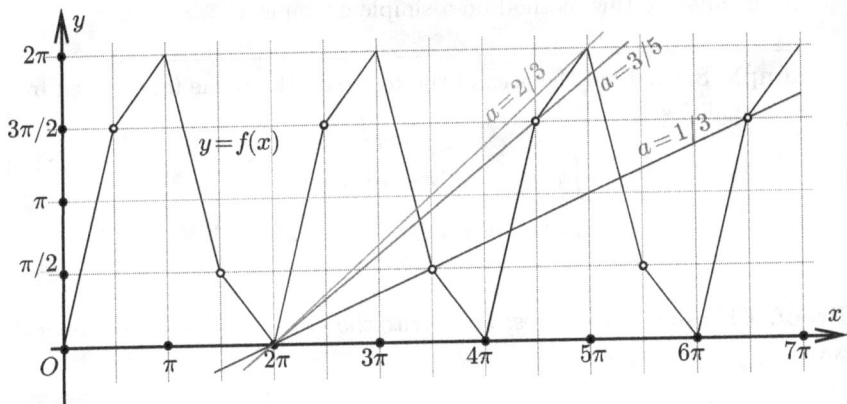

Fig. 8.35. Graphs.

This function is periodic with a period of 2π, and it is linear on each of the sets $[0; \pi/2]$, $[\pi/2; \pi]$, $[\pi; 3\pi/2]$, and $[3\pi/2; 2\pi]$. Since

$$f(0) = f(2\pi) = 0, \quad f(\pi/2) = 3\pi/2, \quad f(\pi) = 2\pi, \quad f(3\pi/2) = \pi/2,$$

we can plot the function on the entire real line.

The graph of the function $y = a(x - 2\pi)$ is a family of lines passing through the point $(2\pi; 0)$.

Next, we select those lines that give three solutions belonging to the set $(-\infty; 7\pi]$. These are the values $a = 1/3$, $a = 2/3$, and $a = 3/5$ (see Fig. 8.35).

Answer: $a = 1/3$, $a = 2/3$, $a = 3/5$. $\qquad\qquad\qquad\qquad\qquad$ □

8.3   The Method of Areas

The area method is a generalization of the interval method. When solving the inequality $f(x)/g(x) \geq 0$, we found the roots of the equations $f(x) = 0$, $g(x) = 0$, thereby splitting the real line into parts in which the sign of the function $f(x)/g(x)$ was preserved. Then, we selected the parts that consisted of the solution to the original problem. When solving the inequality $f(x,y)/g(x,y) \geq 0$ using the domain method, we find all curves on which $f(x,y) = 0$ or $g(x,y) = 0$. They split the plane into parts in which the function $f(x,y)/g(x,y)$ preserves the sign. Then, we select those parts that will give a solution to the original problem.

Let us analyze this method on a simple example

Example 8.12. Find the area of the figure on the plane $(x; y)$ given by the inequalities

$$\begin{cases} x^2 + y^2 - 2x - 4y \leq -1, \\ 3x - 2y + 1 \geq 0. \end{cases}$$

Proof. Choosing full squares, we rewrite the system in a more convenient way:

$$\begin{cases} (x-1)^2 + (y-2)^2 - 4 \leq 0, \\ 3x - 2y + 1 \geq 0. \end{cases} \Longleftrightarrow \begin{cases} f(x,y) \leq 0. \\ g(x,y) \geq 0, \end{cases}$$

where $f(x,y) = (x-1)^2 + (y-2)^2 - 4$, $g(x,y) = 3x - 2y + 1$. The equation $(x-1)^2 + (y2)^2 - 4 = 0$ defines a circle centered at $(1; 2)$ and with a radius of 2. Consequently, the plane is divided by this circle into two parts: external and internal (see Figs. 8.36 and 8.37). Let us check which set satisfies the condition $f(x,y) = (x-1)^2 + (y-2)^2 - 4 \leq 0$. To do this, select an arbitrary point $(1; 2)$ inside the circle and a point $(-2; 2)$ outside the circle. The inequalities fulfilled by them are

$$f(1; 2) = -4 < 0,$$
$$f(-2; 2) = 5 > 0.$$

Therefore, the set lying inside the circle is suitable for us (see Fig. 8.36).

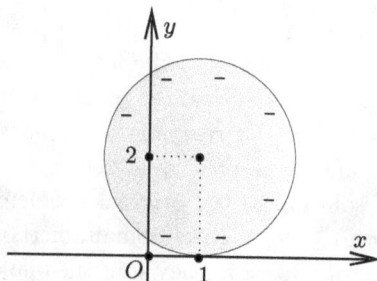

Fig. 8.36. $f(x,y) \leq 0.$

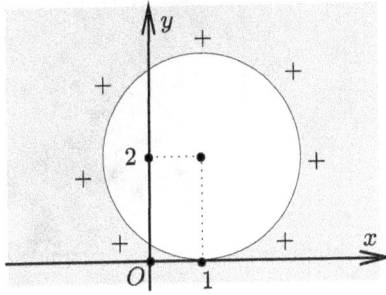

Fig. 8.37. $f(x, y) \geq 0$.

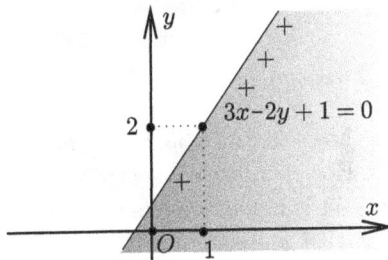

Fig. 8.38. $g(x, y) \geq 0$.

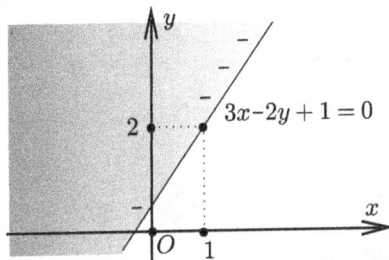

Fig. 8.39. $g(x, y) \leq 0$.

Let's solve the second inequality $g(x, y) = 3x - 2y + 1 \geq 0$. Consider the equation of the line $3x - 2y + 1 = 0$. This straight line divides the plane into two parts (see Figs. 8.38 and 8.39).

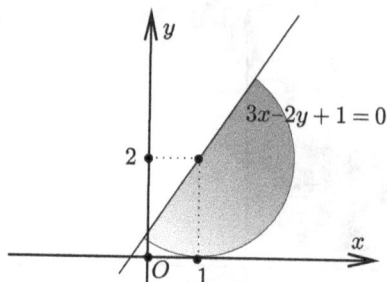

Fig. 8.40. $f(x,y) \leq 0$ and $g(x,y) \geq 0$.

Let us choose a point from each part and determine the one that suits us:

$$g(-2;0) = -5 < 0,$$

$$g(2;0) = 7 > 0.$$

Therefore, the set is as shown in Fig. 8.38. So, we need to find the area of the set shown in Fig. 8.40. However, since the straight line passes through the center of the circle, this set is half of a circle of radius 2. Therefore, the area is equal to $\pi R^2/2 = 2\pi$.

Answer: 2π.

□

Example 8.13. For each value of a, solve the inequality

$$\sqrt{x+2a} > x + \sqrt{2a}.$$

Proof. The inequality is defined only for $a \geq 0$. For convenience, we introduce the notation

$$y = \sqrt{x+2a}, \quad b = \sqrt{2a}.$$

Immediately note that for y and b, the inequalities $y, b \geq 0$ are fulfilled. Since $x = y^2 - b^2$, the initial inequality takes the form $(y^2 - b^2) - (y - b) < 0$, or

$$\begin{cases} (y-b)(y-(1-b)) < 0, \\ y \geq 0, \quad b \geq 0. \end{cases} \tag{8.2}$$

We solve system (8.2) in two ways:

1. **Graphical method.** Since the equations $y = b$ and $y = 1 - b$ define lines on the plane $(b; y)$, it is convenient to depict in the figure the areas of the sign of the function $(y - b)(y - (1 - b))$ (see Fig. 8.41). Given the non-negativity of the variables y and b, we can represent the set that is the solution to the system (see Fig. 8.42).

   It remains to write out the answer. If $b \in [0; 1/2)$, then $y \in (b; 1 - b)$; if $b = 1/2$, then there are no solutions; if $b \in (1/2; 1]$, then $y \in (1 - b; b)$; and if $b > 1$, then $y \in [0; b)$.

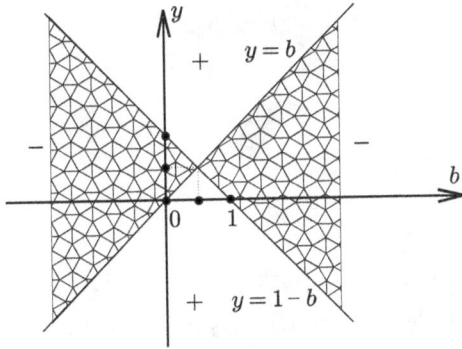

Fig. 8.41. $(y - b)(y - 1 + b) < 0.$

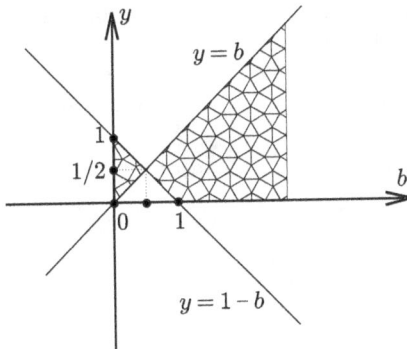

Fig. 8.42. $(y - b)(y - 1 + b) < 0$ and $y_1 b \geq 0.$

2. Let's solve system (8.2) analytically.

To do this, we need to compare the roots of $y_1 = b$, $y_2 = 1 - b$ with each other and with zero. From the equations

$$b = 1 - b, \quad b = 0, \quad 1 - b = 0$$

we find the solutions as $b = 0$, $b = 1/2$, and $b = 1$. Exploring system (8.2) on each of the sets $b \in [0; 1/2)$, $b = 1/2$, $b \in (1/2; 1]$, and $b > 1$ separately, we arrive at the answer in variables $(y; b)$. If $b \in [0; 1/2)$, then $y \in (b; 1 - b)$; if $b = 1/2$, then there are no solutions; if $b \in (1/2; 1]$, then $y \in (1 - b; b)$; and if $b > 1$ then $y \in [0; b)$.

Let us return to the variables (a, x):

$$\begin{cases} b \in [0; 1/2), & y \in (b; 1 - b), \\ b = 1/2, & \text{there are no solutions,} \\ b \in (1/2; 1], & y \in (1 - b; b), \\ b > 1, & y \in [0; b). \end{cases}$$

$$\Longleftrightarrow \begin{cases} \sqrt{2a} \in [0; 1/2], & \sqrt{x + 2a} \in (\sqrt{2a}; 1 - \sqrt{2a}), \\ \sqrt{2a} = 1/2, & \text{there are no solutions,} \\ \sqrt{2a} \in (1/2; 1], & \sqrt{x + 2a} \in (1 - \sqrt{2a}; \sqrt{2a}), \\ \sqrt{2a} > 1, & \sqrt{x + 2a} \in [0; \sqrt{2a}). \end{cases}$$

$$\Longleftrightarrow \begin{cases} a \in [0; 1/8], & x + 2a \in (2a; 1 - 2\sqrt{2a} + 2a), \\ a = 1/8, & \text{there are no solutions,} \\ a \in (1/8; 1/2], & x + 2a \in (1 - 2\sqrt{2a} + 2a; 2a), \\ a > 1/2, & x + 2a \in [0; 2a). \end{cases}$$

$$\Longleftrightarrow \begin{cases} a \in [0; 1/8], & x \in (0; 1 - 2\sqrt{2a}), \\ a = 1/8, & \text{there are no solutions,} \\ a \in (1/8; 1/2], & x \in (1 - 2\sqrt{2a}; 0), \\ a > 1/2, & x \in [-2a; 0). \end{cases}$$

Answer: If $a < 0$ then there are no solutions; if $a \in [0; 1/8)$, then $x \in (0; 1 - 2\sqrt{2a})$; if $a = 1/8$ then there are no solutions; if $a \in (1/8; 1/2]$, then $x \in (1 - 2\sqrt{2a}; 0)$; and if $a > 1/2$, then $x \in [-2a; 0)$. □

Example 8.14. Find all the values of p for which there is a set of solutions to the inequality

$$(p - x^2)(p + x - 2) < 0$$

that does not contain any solution to the inequality $x^2 \leq 1$.

Proof. This example can be solved analytically. But now, we solve it based on a graphical image. We show how it can be solved using the method of areas. In Fig. 8.43, the graphs of the functions $p = x^2$ and $p + x = 2$ are shown. Using the area method, we place the signs of the function standing on the left side of the inequality in the areas formed by these curves. Note (see Fig. 8.44) that for $p \in (0; 3)$, there are solutions to x of the original inequality from the segment $[-1; 1]$, and for the remaining $p \in (-\infty; 0] \cup [3; +\infty)$, there are no solutions to x from the segment $[-1; 1]$.

Answer: $p \in (-\infty; 0] \cup [3; +\infty)$. □

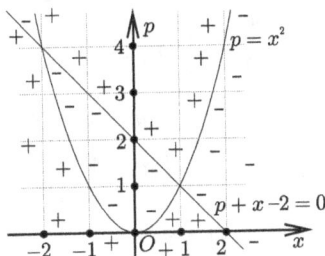

Fig. 8.43.   Method of areas.

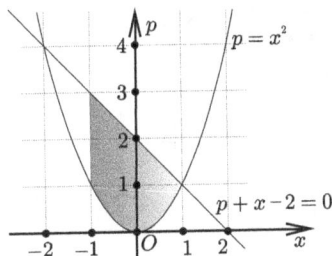

Fig. 8.44. $(p - x^2)(p + x - 2) < 0$ and $x^2 \leq 1$.

Example 8.15. Prove that the set given on the coordinate plane by the condition

$$|3x + 6| + |2y + 3x - 2| < 6$$

is a parallelogram centered at the intersection point of the lines $3x+6 = 0$ and $2y + 3x - 2 = 0$, which are the diagonals of this parallelogram. Find the area of the parallelogram.

Proof. Note that

$$|a| + |b| < c \Longleftrightarrow \begin{cases} |a + b| < c, \\ |a - b| < c. \end{cases}$$

Indeed, this inequality is easy to check by considering all possible combinations of the signs of the numbers a and b. Using this remark, we find that the initial inequality is equivalent to the system

$$|2y + 3x - 2| + |3x + 6| < 6 \Longleftrightarrow \begin{cases} |(2y + 3x - 2) + (3x + 6)| < 6, \\ |(2y + 3x - 2) - (3x + 6)| < 6. \end{cases}$$

$$\Longleftrightarrow \begin{cases} |2y + 6x + 4| < 6, \\ |2y - 8| < 6. \end{cases} \Longleftrightarrow \begin{cases} |y + 3x + 2| < 3, \\ |y - 4| < 3. \end{cases}$$

Solving the inequality $|y + 3x + 2| < 3$, we get the set $-5 < y + 3x < 1$ (see Fig. 8.45).

Solving the inequality $|y - 4| < 3$, we get the set $1 < y < 7$ (see Fig. 8.46).

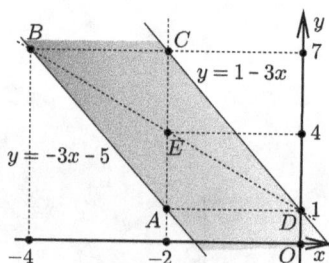

Fig. 8.45.   The set $-5 < y + 3x < 1$.

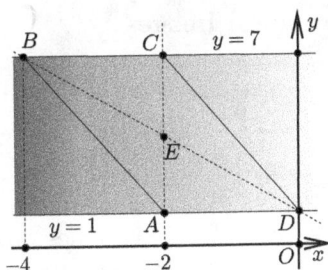

Fig. 8.46. The set $1 < y < 7$.

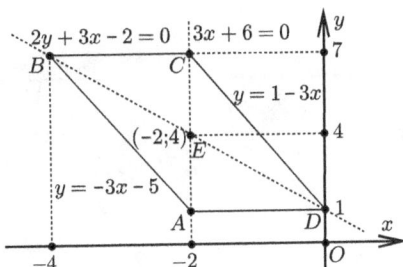

Fig. 8.47. The set of intexsection $-5cy + 3x < 1$ and $1 < y < 7$.

The intersection of the sets $-5 < y + 3x < 1$ and $1 < y < 7$ (see Fig. 8.47) is indeed a parallelogram with sides

$$AB: \quad y + 3x + 5 = 0,$$

$$BC: \quad y = 7,$$

$$CD: \quad y + 3x - 1 = 0,$$

$$DA: \quad y = 1.$$

Therefore, the diagonals of this parallelogram are the lines AC: $3x + 6 = 0$ and BD: $2y + 3x - 2 = 0$, which intersect at the point $(-2; 4)$. Find the area of the parallelogram

$$S_{ABCD} = AC \cdot AD = 2 \cdot 6 = 12.$$

Answer: $S_{ABCD} = 12$. $\qquad\square$

8.4   Tasks Using Graphical Images

> **Example 8.16.** Find the smallest value of the expression
>
> $$\sqrt{(x-9)^2+4}+\sqrt{x^2+y^2}+\sqrt{(y-3)^2+9}.$$

Proof. The solution to this problem becomes obvious when looking at the drawings in Figs. 8.48 and 8.49. Denote $d_1 = \sqrt{x^2+y^2}$, $d_2 = \sqrt{(y-3)^2+9}$, and $d_3 = \sqrt{(x-9)^2+4}$. The point $O(0;0)$ is the origin of coordinates. The original expression is the sum of the distances between the three points $A(x;y)$, $B(x+3;3)$, and $C(12;5)$. In fact,

$$OA = \sqrt{(x-0)^2+(y-0)^2} = d_1,$$

$$AB = \sqrt{(x+3-x)^2+(3-y)^2} = \sqrt{(y-3)^2+9} = d_2,$$

$$BC = \sqrt{(12-(x+3))^2+(5-3)^2} = \sqrt{(x-9)^2+4} = d_3.$$

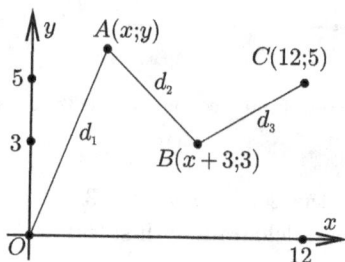

Fig. 8.48. $d_1 + d_2 + d_3$.

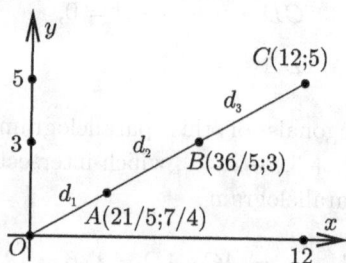

Fig. 8.49.   The smallest $d_1 + d_2 + d_3$.

Therefore, the smallest value of the sum of the distances d_1, d_2, and d_3 will be reached if the points A and B are on the same segment connecting the points O and C (see Fig. 8.49).

Let's check that such an arrangement of points is possible. The equation of a straight line passing through the points O and C is $x/12 = y/5$; however, since this line must pass through the point $B(x+3;3)$, we arrive at the system

$$\begin{cases} x/12 = y/5, \\ (x+3)/12 = 3/5. \end{cases} \Longleftrightarrow \begin{cases} x = 21/5, \\ y = 7/4. \end{cases}$$

Thus, we proved that an arrangement where all points are on the same straight line is possible. Therefore, the smallest value of the expression is $\sqrt{12^2 + 5^2} = 13$.

Answer: 13. $\qquad\square$

Example 8.17. For each value of a, solve the system of equations

$$\begin{cases} 2^{1+x} = 32a\sqrt{2}, \\ \sqrt{x^2 + a^2 + 2 - 2x - 2a} + \sqrt{x^2 + a^2 - 6x + 9} = \sqrt{5}. \end{cases}$$

Proof. Let us write the second equation in the form

$$\sqrt{(x-1)^2 + (a-1)^2} + \sqrt{(x-3)^2 + a^2} = \sqrt{5}.$$

This equation means that the sum of the distances from point $(x;a)$ to points $(1;1)$ and $(3;0)$ is equal to $\sqrt{5}$. Since the distance between the points $(1;1)$ and $(3;0)$ is also equal to $\sqrt{5}$, this means that the point $(x;a)$ must lie on the segment connecting the points $(1;1)$ and $(3;0)$ (see the Figures 8.50–8.52). In other words, it satisfies the equation $a = (3-x)/2$ and the condition $x \in [1;3]$. Thus, the original system is equivalent to the system

$$\begin{cases} 2^{1+x} = 32a\sqrt{2}, \\ 2a = 3 - x, \quad x \in [1;3]. \end{cases}$$

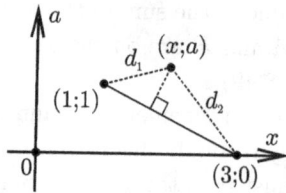

Fig. 8.50. $d_1 + d_2 > \sqrt{5}$.

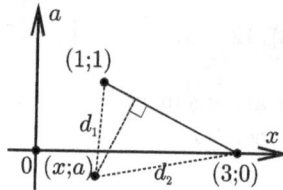

Fig. 8.51. $d_1 + d_2 > \sqrt{5}$.

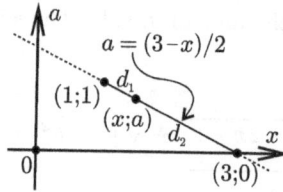

Fig. 8.52. $d_1 + d_2 = \sqrt{5}$.

Substituting $2a$ into the first equation, we get

$$2^{1+x} = 16(3-x)\sqrt{2} \iff 2^{x-7/2} = 3-x$$
$$\iff 2^{x-7/2} + x = 3.$$

Since the function $2^{x-7/2} + x$ is increasing (as the sum of two increasing functions), the equation takes no more than one solution. By selecting, we find the solution as $x = 5/2$; it is the only one, and it corresponds to $a = 1/4$.

Answer: If $a = 1/4$, then $x = 5/2$; for other values of a, there are no solutions.   □

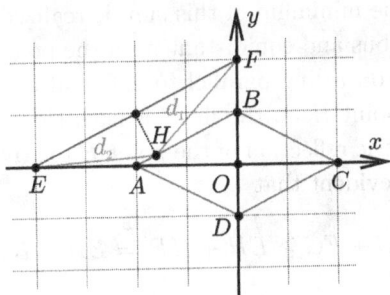

Fig. 8.53. Geometric interpritation.

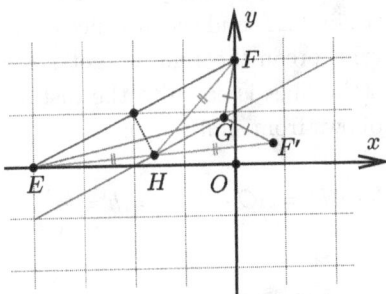

Fig. 8.54. F' is symmetric to F.

Example 8.18. Find the minimum value of the expression

$$\sqrt{(x+6)^2 + y^2} + \sqrt{x^2 + (y-4)^2}$$

provided $2|x| + 3|y| = 6$.

Proof. Since the expression $\sqrt{(x+6)^2 + y^2} + \sqrt{x^2 + (y-4)^2}$ is equal to the sum of the distances from the point $(x; y)$ to the points with coordinates $(-6; 0)$ and $(0; 4)$, and the geometric shape of the points $2|x| + 3|y| = 6$ on the plane is a rhombus with vertices $(3; 0)$, $(0; -2)$, $(-3; 0)$, and $(0; 2)$, then the problem is reduced to studying the sum of distances from the point $(x; y)$ lying on the rhombus $2|x| + 3|y| = 6$ to points with coordinates $(-6; 0)$ and $(0; 4)$ (see Fig. 8.53).

We prove that the minimum of this sum is realized at a point lying on the side of the rhombus and equidistant from the points $(-6;0)$ and $(0;4)$. Let the point G lie on a line parallel to EF and from the line EF at a distance of h. The point H on the line l is such that $EH = HF$, and the point F' is a symmetric reflection of the point F relative to the line l. Then (see Fig. 8.54), it is evident that

$$EG + FG \geq EH + HF' = EH + HF.$$

In our case, the side of the rhombus AB is parallel to EF, and the point H of the straight line AB is such that $EH = FH$ lies on the side of the rhombus. Therefore, according to the proven, all other points of the rhombus give the total distance to the points E and F to be greater than $EH+FH$. It remains to find EF and the distance between the lines EF and AB. Applying the Pythagorean theorem, we get $EF = \sqrt{16+36} = 2\sqrt{13}$. The distance between the lines is equal to the distance from the origin to the line AB (which follows from similarity):

$$h \cdot AB = AO \cdot BO \Longrightarrow h = \frac{6}{\sqrt{13}}.$$

So,

$$EH + HF = 2\sqrt{\frac{36}{13} + 13} = 2\sqrt{\frac{205}{13}}.$$

Answer: $2\sqrt{\frac{205}{13}}$. □

Remark 8.1.

- The justification that the minimum of the sum is realized at a point lying on the side of the rhombus and equidistant from the points $(-6;0)$ and $(0;4)$ can be found using the convexity of the function $f(x) = \sqrt{x}$. Indeed, the estimate will be obtained from $f\left(\frac{a+b}{2}\right) \geq \frac{1}{2}(f(a) + f(b))$.
- The distance between parallel lines $\frac{x}{-3} + \frac{y}{2} = 1$ and $\frac{x}{-3} + \frac{y}{2} = 2$ can be found immediately from the formula $\varrho = \frac{|2-1|}{\sqrt{1/9+1/4}} = \frac{6}{\sqrt{13}}$.
- The point H has coordinates $(27/13; 8/13)$.

Example 8.19. Find all the values of the parameters a and b for which the system of equations

$$\begin{cases} x^2 + y^2 + 5 = b^2 + 2x - 4y, \\ x^2 + (12 - 2a)x + y^2 = 2ay + 12a - 2a^2 - 27 \end{cases}$$

has exactly two different solutions $(x_1; y_1)$ and $(x_2; y_2)$ satisfying the condition

$$\frac{x_1 + x_2}{y_2 - y_1} = \frac{y_1 + y_2}{x_1 - x_2}.$$

Proof. The last condition means that $(x_1; y_1)$ and $(x_2; y_2)$ belong to a circle centered at $(0; 0)$ because

$$\frac{x_1 + x_2}{y_2 - y_1} = \frac{y_1 + y_2}{x_1 - x_2}. \iff \begin{cases} x_1^2 + y_1^2 = x_2^2 + y_2^2, \\ x_1 \neq x_2, \quad y_1 \neq y_2. \end{cases}$$

Selecting the full squares, we rewrite the original system in the following form:

$$\begin{cases} (x - 1)^2 + (y + 2)^2 = b^2, \\ (x + (6 - a))^2 + (y - a)^2 = 9. \end{cases}$$

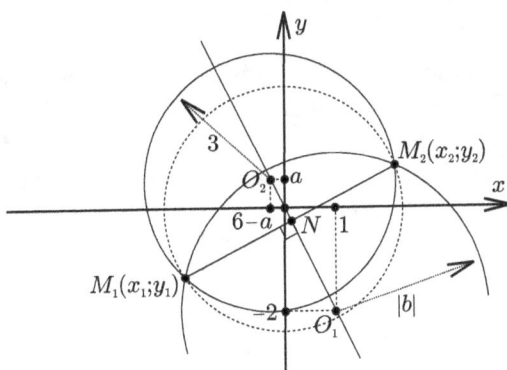

Fig. 8.55.   Three circles.

These equations describe circles with centers at points $O_1(1; -2)$ and $O_2(a - 6; a)$, respectively (see Fig. 8.55). Thus, the coordinates of the points are $M_1(x_1; y_1)$ and $M_2(x_2; y_2)$, which must satisfy the equations of three circles simultaneously.

Let us first consider the circles with centers at points $O_1(1; -2)$ and $O(0; 0)$. Let N be the intersection point of $M_1 M_2$ with OO_1. Since OO_1 is the median perpendicular to $M_1 M_2$, then the points O, N, and O_1 lie on one straight line perpendicular to $M_1 M_2$. Similarly, considering the circles with centers at points $O_2(a - 6; a)$ and $O(0; 0)$, we find that the points O, N, and O_2 lie on one straight line perpendicular to $M_1 M_2$.

Thus, the points O, O_1, and O_2 lie on one straight line passing through the point N. Let's write the equation of a straight line[1] passing through the points O_1 and O_2:

$$\frac{x - 1}{a - 6 - 1} = \frac{y + 2}{a + 2}.$$

Substituting the direct coordinates of the point $O(0; 0)$ into the equation, we get

$$(-1)(a + 2) = 2(a - 7) \Longleftrightarrow 3a = 12 \Longleftrightarrow a = 4.$$

Thus, the equation of a straight line passing through the points O_1 and O_2 can be written as

$$\frac{x - 1}{-3} = \frac{y + 2}{6} \Longleftrightarrow y + 2x = 0.$$

Let's check the fulfilment of the conditions $x_1 \neq x_2$ and $y_1 \neq y_2$ (for the found $a = 4$) arising from the domain of definition in this example. If the equality $x_1 = x_2$ (or $y_1 = y_2$) were true, then the line $M_1 M_2$ would be parallel to the axis Oy (or Ox), but this is not the case since the line $M_1 M_2$ is perpendicular to the line $y + 2x = 0$. (The equation of the line $M_1 M_2$ has the form $x - 2y = -1 - b^2/6$.)

---

[1]The equation of a straight line passing through two given points, (x_1, y_1), (x_2, y_2), is written as

$$\frac{x - x_1}{x_2 - x_1} = \frac{y - y_1}{y_2 - y_1}.$$

In the case of $x_2 = x_1$, it takes the form $x = x_1$, and in the case of $y_2 = y_1$, it takes the form $y = y_1$.

The circles with centers at points O_1 and O_2 intersect at two points if and only if the following conditions are fulfilled:

$$|R_2 - R_1| < O_1O_2 < R_1 + R_2,$$

$$\Longleftrightarrow ||b| - 3| < O_1O_2 < |b| + 3.$$

Since $O_1O_2 = \sqrt{(1+2)^2 + (-2-4)^2} = \sqrt{45}$, then $\sqrt{45}-3 < |b| < \sqrt{45}+3$.

Answer: $a = 4$, $\sqrt{45} - 3 < |b| < \sqrt{45} + 3$. $\square$

Problems

8.1. Find all the values of a for which the system

$$\begin{cases} x^2 + y^2 = 1, \\ y - |x| = a \end{cases}$$

has exactly two different solutions.

8.2. Find all the values of a for which the system

$$\begin{cases} (|x| - 5)^2 + (y - 4)^2 = 9, \\ (x - 2)^2 + y^2 = a^2 \end{cases}$$

has a single solution.

8.3. Find all the values of a for which the equation

$$(a + 2 - x^2)(||x - 1| - 1| - a) = 0$$

has exactly five different solutions.

8.4. Find the smallest value of the expression $a^2 + (b - 1)^2$ on the set of such numbers a and b for which the equation

$$||x - 4| - 2| - ax + (4a - b) = 0$$

has exactly three distinct roots. Indicate at which a and b this lowest value is reached.

8.5. For what values of a does the equation

$$2|x - 2a| - a^2 + 15 + x = 0$$

have no solutions? For what values of a does the equation have solutions and all solutions belong to the segment $[-9; 10]$?

8.6. Find all the values of a for which the equation

$$||x + a| - 2x| - 3x = 7|x - 1|$$

has at most one root.

8.7. Find all the values of a, for each of which the smallest value of the function $f(x) = 2ax + |x^2 - 6x + 8|$ is less than 1.

8.8. Find all the values of a for which the equation $|x^2 - 6x + 8| + |x^2 - 6x + 5| = a$ has exactly three different solutions.

8.9. Find all values of a for which the equation

$$||x - a| + 2x| + 4x = 8|x + 1|$$

has no root.

8.10. Find all the values of a for which the function $f(x) = x^2 - 3|x - a^2| - 5x$ has at least one maximum point.

8.11. For what values of a there is a single solution to the system

$$\begin{cases} x^2 + y^2 = 9, \\ (x - 4)^2 + (y - 3)^2 = a? \end{cases}$$

8.12. Find all the values of a for which the function

$$f(x) = x^2 - 4|x - a^2| - 8x$$

has more than two extremum points.

8.13. Find all the values of a for each of which the system

$$\begin{cases} (x + 2)^2 + y^2 = a, \\ |x| + |y - 1| = 1 \end{cases}$$

has a single solution.

8.14. Find all the values of a for each of which the equation

$$ax + \sqrt{-7 - 8x - x^2} = 2a + 3$$

has a single root.

8.15. Find all the values of a for which the system

$$\begin{cases} y = 2|x| + |x - 7|, \\ y = 2|x - 3| + x + a \end{cases}$$

has a single solution. Indicate this solution.

8.16. Find all the values of a for which the equation

$$a - |x - 1| - |x - 2| - |x - 3| = 2|x + 1| + |x + 2|$$

have infinitely many solutions.

8.17. Find all the values of a for which the equation

$$4|x - a| + a - 2 - 2x = 0$$

has solutions and all solutions belong to the segment $[-2; 1]$.

8.18. Find all the values of a for which the equation

$$(13 + a - 6x + x^2) \cdot (a + 5 - |x - 3|) = 0$$

has exactly three different solutions.

8.19. For each value of a, specify the number of common points of the graphs of the functions

$$y = x^2 - 4x + |4 - 2x|$$

and $y = a$. Indicate the coordinates of these points.

8.20. For what values of a does the system

$$\begin{cases} y^2 + 2(x - 2)y + (x^2 - 4)(2x - x^2) = 0, \\ y = a(x - 4) \end{cases}$$

have exactly three different solutions?

8.21. Find all the values of a for which the system

$$\begin{cases} x^2 + 8x + y^2 + 8y + 23 = 0, \\ x^2 - 2a(x + y) + y^2 + a^2 = 0 \end{cases}$$

has a single solution.

8.22. Find all the values of a for which the system

$$\begin{cases} |2x + 3y| + |2x - 3y| = 7, \\ x^2 + y^2 = a^2 - 4 - 4y \end{cases}$$

has at least one solution.

8.23. Find all the values of a for which the system

$$\begin{cases} ax^2 + 4ax - y + 7a + 2 = 0, \\ ay^2 - x - 2ay + 4a - 1 = 0 \end{cases}$$

has a single solution.

8.24. Find all the values of a for which the system

$$\begin{cases} (x-a)^2 + 3y^2 - 2y = 0, \\ |x| - y = 4/3 \end{cases}$$

has a single solution.

8.25. Find all the values of a for which the equation $2|x + 1| = 8 - |8(x-a)^2 - |x+1| - 14|$ has exactly three different solutions.

8.26. Find all the values of a for which the system

$$\begin{cases} (x-4)^2 + (y-6)^2 = 25, \\ y = |x-a| + 1 \end{cases}$$

has exactly three different solutions.

8.27. Find all the values of a for each of which there are many solutions to the inequality

$$\sqrt{3-x} + |x-a| \le 2$$

is a segment.

Tasks for Section 8.2

8.28. Find all the values of a for which the system

$$\begin{cases} (y^2 + |x| + |2 - x| - 2) \cdot (xy - x + y - 2) = 0, \\ x - 2a - 1 + (y-1)(a+1)^2 = 0 \end{cases}$$

has exactly two different solutions.

8.29. Find all the values of a for which the system

$$\begin{cases} |a|^{y^2} = \sqrt[7]{-4x^2 + 24x - 32}, \\ y = 4x^2 - 24x + 32 \end{cases}$$

has at least two solutions.

8.30. Find all the values of a for which the system

$$\begin{cases} (x - y^2 + 1)(y - \sqrt{6}|x|) = 0, \\ 2ay - x = 1 + a^2 \end{cases}$$

has exactly two different solutions.

8.31. Find all the values of a for which the system

$$\begin{cases} |a|^{2x-7y+12} = e^2(2x - 6y + 11), \\ 4y - x = 6 \end{cases}$$

has exactly two different solutions.

8.32. Find all the values of a for which the equation

$$|x^2 - 6x + 8| + 2 = \log_a x$$

has a unique solution.

8.33. Solve the equation

$$x^2 = \arcsin(\sin x) + 10x.$$

8.34. Consider the functions

$$f(x, y) = |y| + 2|x| - 2$$

and

$$g(x, y, a) = x^2 + (y - a)(y + a).$$

(a) Find the smallest positive value of a at which the system of equations

$$\begin{cases} f(x, y) = 0, \\ g(x, y, a) = 0 \end{cases}$$

has exactly four different solutions?

(b) With this value of a, find the area of the figure, coordinates (x, y) of all points of which satisfy the inequality

$$\frac{f(x, y)}{g(x, y, a)} \le 0.$$

8.35. Solve the inequality

$$2\arcsin(\sin x) + \arccos(\cos x) \ge -x - 3.$$

8.36. Find all positive a for which the equation

$$(4\pi a + \arcsin(\sin x) + 3\arccos(\cos x) - ax) \cdot (2 + \tan^2 x)^{-1} = 0$$

has exactly three different solutions.

8.37. Find all the values of a for which the roots of the equation

$$\sqrt{x + 3 - 4\sqrt{x - 1}} + \sqrt{x + 8 - 6\sqrt{x - 1}} = a$$

exist and belong to the segment $[2; 17]$.

8.38. Find all the values of a and n for which the difference between the largest and smallest positive roots of the equation

$$\underbrace{||\cdots|||}_{n \text{ times}} x - 1| - 1| - 1| - \cdots - 1| - 1| = a$$

is equal to 18.3.

8.39. Find all the values of a for which the equation

$$a + \sqrt{6x - x^2 - 8} = 3 + \sqrt{1 + 2ax - a^2 - x^2}$$

has exactly one solution.

8.40. Find all the values of a and b for which the system of equations

$$\begin{cases} x^2 + y^2 + 5 = b^2 + 2x - 4y, \\ x^2 + (12 - 2a)x + y^2 = 2ay + 12a - 2a^2 - 27 \end{cases}$$

has two solutions $(x_1; y_1)$ and $(x_2; y_2)$ satisfying the condition

$$\frac{x_1 - x_2}{y_2 - y_1} = \frac{y_1 + y_2}{x_1 + x_2}.$$

8.41. Find all the values of a for which the system

$$\begin{cases} (xy - y - 9)(y + x^2 - 1) = 0, \\ y = a(x - 3). \end{cases}$$

has exactly three different solutions.

8.42. Find all the values of a for which the equation

$$\sin \arccos(5x) = a + \arcsin \sin(7x - 3)$$

has a unique solution.

8.43. Find all the values of a for which the graphs of the functions

$$y = \frac{3x + 1}{x} \quad \text{and} \quad y = \frac{4x + 3a - 7}{ax - 1}$$

divide the coordinate plane into exactly five parts.

Tasks for Section 8.3

8.44. For what values of a on the plane there is a circle containing all points satisfying the system of ine- qualities:

$$\begin{cases} 2y - x \leq 1, \\ y + 2x \leq 2, \\ y + ax \geq -1? \end{cases}$$

8.45. Find all the values of a for which the system

$$\begin{cases} x^2 + a \leq -2x, \\ a + 2 + x \geq 0 \end{cases}$$

has a unique solution.

8.46. Six numbers form an increasing arithmetic progression. The first, second, and fourth terms of this progression are solutions to the inequality

$$\log_{0,5x-1}\left(\log_4\left(\frac{x-11}{x-8}\right)\right) \geq 0,$$

and the rest are not solutions to this inequality. Find the set of all possible values of the first term of such progressions.

8.47. Find the area of the figure whose point coordinates satisfy the relations

$$\begin{cases} x^2 + y^2 \leq 9, \\ y + 1 \geq 0, \\ 3y + 6 \geq 2|x|. \end{cases}$$

8.48. Find the area of the figure located on the coordinate plane and consisting of points (x, y) satisfying the inequality

$$\log_{(x^2+y^2)/2}(x - y) > 1.$$

8.49. Find the perimeter of the figure whose points on the coordinate plane $(x; y)$ satisfy the system

$$\begin{cases} y > ||x - 2| - 1|, \\ x^2 + y^2 < 4x + 2y - 3. \end{cases}$$

8.50. Find the area of the shape defined on the coordinate plane by the conditions

$$\begin{cases} y \leq \sqrt{4 - x^2}, \\ y \geq |x - 1| - 3. \end{cases}$$

8.51. Find the area of the figure located on the coordinate plane and consisting of points (x, y) satisfying the inequality

$$x^2 + y^2 \leq 6|x| - 6|y|.$$

8.52. Find all the values that the sum of $x + a$ can take, provided

$$|2x + 4 - 2a| + |x - 2 + a| \leq 3.$$

8.53. Find the area of the figure given on the coordinate plane by the ratio

$$|y - x^2/2| + |y + x^2/2| \leq 2 + x.$$

8.54. Find all the values of a for which the system

$$\begin{cases} |x + 2y + 1| \leq 11, \\ (x - a)^2 + (y - 2a)^2 = 2 + a \end{cases}$$

has a unique solution.

8.55. Find all the values of a for which there is exactly one solution to the inequality

$$x^2 + (5a + 3)x + 4a^2 \leq 4$$

that satisfies the inequality

$$ax(x - 4 - a) \leq 0.$$

8.56. Find all the values of a for which the system of inequalities

$$\begin{cases} \sqrt{(x - 2a)^2 + (y - a)^2} \leq \dfrac{|a|}{6\sqrt{5}}, \\ x - 2y \geq 1 \end{cases}$$

has solutions.

8.57. Find all positive values of a for which the system of inequalities

$$\begin{cases} (x+y+a)^2 + (x-y-a)^2 \le (a-1)^2, \\ (x+y-2a)^2 + (x-y+3a)^2 \le (8a-5)^2 \end{cases}$$

has no solutions.

8.58. For what values p is the area of a figure defined on the coordinate plane by the equation

$$|2x+y| + |x-y+3| \le p$$

will be equal to 24?

8.59. Find all the values of a for which the system

$$\begin{cases} x^2 + y^2 - 6|x| - 6|y| + 17 \le 0, \\ x^2 + y^2 - 2y = a^2 - 1 \end{cases}$$

has at least one solution.

8.60. Obtain the equation of the circle with the smallest radius inside which the set given on the coordinate plane by the following condition is placed:

$$|y - 2x - 1| + |2x - 4| < 4.$$

8.61. For what values of a does the system

$$\begin{cases} x^2 + (2-3a)x + 2a^2 - 2a < 0, \\ ax = 1 \end{cases}$$

have solutions?

8.62. For each value of $a \ge 0$, solve the inequality

$$\frac{x^2(x-2)}{x+2} + ax^2 + \frac{ax}{x+2} - 2ax + a^2 \ge 0.$$

8.63. Find all the values of c for each of which the set of points of the plane whose coordinates $(x; y)$ satisfy the system

$$\begin{cases} \dfrac{x^2 + y^2 - 16x + 10y + 65}{x^2 + y^2 - 14x + 12y + 79} \le 0, \\ (x-c)(y+c) = 0, \end{cases}$$

are a segment.

8.64. Find all the values of a for which the system

$$\begin{cases} \dfrac{x - ax - a}{x - 2 + 2a} \geq 0, \\ x - 8 > ax \end{cases}$$

has no solutions.

8.65. For each value of b, solve the inequality

$$\sqrt{x + 4b^2} > x + 2|b|.$$

8.66. For each value of a belonging to the segment $[-1, 0]$, solve the inequality

$$\log_{x+a}(x^2 - (a + 1)x + a) \geq 1.$$

Tasks for Section 8.4

8.67. Find the smallest value of the expression

$$\sqrt{(x - 6)^2 + 36} + \sqrt{x^2 + y^2} + \sqrt{(y - 6)^2 + 9}.$$

8.68. For each valid value of a, solve the system

$$\begin{cases} \sqrt{x^2 + a^2 - 2x - 22a + 122} = 2\sqrt{37} - \sqrt{x^2 + a^2 + 2x + 2a + 2}, \\ \log_{x+1} 4 + \log_a 4 = 0. \end{cases}$$

8.69. Find the minimum value of the expression

$$\sqrt{(x - 3)^2 + y^2} + \sqrt{x^2 + (y + 6)^2},$$

provided $2|x| + |y| = 4$.

8.70. For each value of a, solve the system

$$\begin{cases} x^2 + a^2 - 14x - 10a + 58 = 0, \\ \sqrt{x^2 + a^2 - 16x - 12a + 100} \\ \quad + \sqrt{x^2 + a^2 + 4x - 20a + 104} = 2\sqrt{29}. \end{cases}$$

8.71. Solve the system of equations

$$\begin{cases} 2^{2-x} = 4y\sqrt{2}, \\ \sqrt{x^2 + y^2 + 1 - 2x} + \sqrt{x^2 + y^2 - 6x - 2y + 10} = \sqrt{5}. \end{cases}$$

8.72. For what values of a does the system

$$\begin{cases} y^2 - (2a+1)y + a^2 + a - 2 = 0, \\ \sqrt{(x-a)^2 + y^2} + \sqrt{(x-a)^2 + (y-3)^2} = 3 \end{cases}$$

have a single solution?

8.73. Find all the values of a and b for which the system of equations

$$\begin{cases} x^2 + 40 - a^2 = 4y - y^2 - 12x, \\ x^2 + y^2 + (-2b-8)x = 2by - 2b^2 - 8b \end{cases}$$

has exactly two different solutions $(x_1; y_1)$ and $(x_2; y_2)$ satisfying the condition

$$\frac{y_1 + y_2}{x_1 - x_2} = \frac{x_1 + x_2}{y_2 - y_1}.$$

Hints and Answers

8.1. $a \in \{-\sqrt{2}\} \cup (-1; 1)$.

8.2. $a = 2$, $a = 3 + \sqrt{65}$.

8.3. $a = 1$, $a = (5 - \sqrt{17})/2$.

8.4. The smallest value of $1/5$ is achieved when $a = \pm 2/5$, $b = 4/5$.

8.5. (a) $a \in (-3; 5)$, (b) $a \in [2 - 2\sqrt{7}; -3] \cup \{5\}$.

8.6. $a \in [-6; 4]$.

8.7. $a \in (-\infty; 1/4) \cup (3 + \sqrt{7}; +\infty)$.

8.8. $a = 5$.

8.9. $a \in (-7; 5)$.

8.10. $a \in (-2; -1) \cup (1; 2)$.

8.11. $a = 4$, 64.

8.12. $a \in (-\sqrt{6}; -\sqrt{2}) \cup (\sqrt{2}; \sqrt{6})$.

8.13. $a = 2$, $a = 10$.

8.14. $a \in [-1; -1/3) \cup \{0\}$.

8.15. If $a \in (-1; 1)$, then $(x; y) = ((13 - a)/2; (27 - a)/2)$; if $a > 7$, then $(x; y) = ((1 - a)/2; (3a + 11)/2)$.

8.16. $a = 10$.

8.17. $a \in [-2/5; 2/3]$.

8.18. $a = -5$, $a = -4$.

8.19. If $a < -4$, then there are no common points; if $a = -4$, then $(x; y) = (2; -4)$; if $a > -4$ then $(x; y) = (3 - \sqrt{5 - a}; a)$ and $(x; y) = (1 + \sqrt{5 - a}; a)$.

8.20. $a = -3/5$; 0; $-8 \pm 4\sqrt{3}$; $6 \pm 4\sqrt{2}$.

8.21. $-5 \pm \sqrt{2}$; $-11 \pm 7\sqrt{2}$.

8.22. $a \in [-\sqrt{1885}/12; -5/6] \cup [5/6; \sqrt{1885}/12]$.

8.23. $a \in \{-1/2; 0; 1/6\}$.

8.24. $a = \pm 1$, $\pm 7/3$.

8.25. $a = -7/2$, $3/2$.

8.26. 4; $5\sqrt{2} - 1$; $9 - 5\sqrt{2}$.

8.27. $a \in (-1; 1) \cup [5/4; 5)$.

8.28. $a \in [-4; -2 - \sqrt{2}] \cup [-2 + \sqrt{2}; 0]$.

8.29. $a \in (-e^{1/(14e)}; 0) \cup (0; e^{1/(14e)})$.

8.30. $a \in \{\sqrt{6}/3; \sqrt{6}/2\} \cup (-1/(2\sqrt{6}); 1/(2\sqrt{6})]$.

8.31. $a \in (-e^2; -1) \cup (1; e^2)$.

8.32. $a \in (0; 1) \cup \{2\}$.

8.33. $0, (9 + \sqrt{81 + 12\pi})/2$.

8.34. a) $2/\sqrt{5}$; b) $4 - 4\pi/5$.

8.35. $x \in [-3/4 - 3\pi/2; -\pi + 3/2] \cup [-3/2; +\infty)$.

8.36. $a = 1, a = 2/3, a = 4/5$.

8.37. $a \in [1; 3]$.

8.38. $n = 19, a = 0, 15$.

8.39. $a \in [2; 3) \cup (3; 4]$.

8.40. $a = 4, \sqrt{45} - 3 < |b| < \sqrt{45} + 3$.

8.41. $a = -6 - 4\sqrt{2}, a = 3/5$.

8.42. $a \in [\pi - 22/5; \pi - 8/5) \cup \{\pi - 3 + \sqrt{74}/5\}$.

8.43. $a \in [0; 1]$.

8.44. $a \in (-1/2; 2)$.

8.45. $-3; 1$.

8.46. $a_1 \in (2; 2.5)$. **Hint:** Write the inequalities for a_1 and d, and solve them using the domain method.

8.47. $9(\pi + 1)/2$.

8.48. $S = \pi/3 + 2\sqrt{3}$.

8.49. $P = 3\pi/\sqrt{2} + 2\sqrt{2}$.

8.50. $S = 2\pi + 7$.

8.51. $10\pi + 16$.

8.52. $[-1; 5]$.

8.53. $S = 15/2$.

8.54. $a = -2, a = 3$.

8.55. $a \in \{-5/3, -3/2, -1, 1\}$.

8.56. $a \in (-\infty; -6] \cup [6; +\infty)$.

8.57. $a \in (3/7; 3/2)$.

8.58. $p = 6$.

8.59. $a \in [-6; 1 - \sqrt{13}] \cup [\sqrt{13} - 1; 6]$.

8.60. $(x - 2)^2 + (y - 5)^2 = 20.$

8.61. $a \in (-1; (1 - \sqrt{3})/2) \cup (1; (1 + \sqrt{3})/2).$

8.62. If $a \in [0; 1)$, then $x \in (-\infty; -2) \cup [-2a/(a + 1); 1 - \sqrt{1 - a}] \cup [1 + \sqrt{1 - a}; +\infty)$; if $a \geq 1$, then $x \in (-\infty; -2) \cup [-2a/(a + 1); +\infty).$

8.63. $(5 - 2\sqrt{6}; 8 - 2\sqrt{6}) \cup (5 + 2\sqrt{6}; 8 + 2\sqrt{6}).$

8.64. $[1; 3].$

8.65. If $|b| \in [0; 1/4)$, then $x \in (0; 1 - 4|b|)$; if $|b| = 1/4$, then there are no solutions; if $|b| \in (1/4; 1/2]$, then $x \in (1 - 4|b|; 0)$; if $|b| > 1/2$, then $x \in [-4b^2; 0).$

8.66. If $a \in (-1; 0)$, then there are no solutions; if $a = -1$, then $x \in (2; +\infty)$; if $-1 < a < -1/2$, then $x \in (1; a + 2] \cup (1 - a; +\infty)$; if $a = -1/2$, then $x \in (1; 3/2) \cup (3/2; +\infty)$; if $-1/2 < a < 0$, then $x \in (1; 1 - a) \cup [a + 2; +\infty)$; if $a = 0$, then $x \in [2; +\infty).$

8.67. 15.

8.68. If $a = 2$, then $x = -1/2$; there are no solutions for the remaining a.

8.69. $x \in (2; 3].$

8.70. If $a = (180 + 2\sqrt{415})/29$, then $x = (217 - 5\sqrt{415})/29$; there are no solutions for the remaining a.

8.71. $(3/2; 1/4).$

8.72. $a \in [-2; 1) \cup (1; 4].$

8.73. $b = -1$, $\sqrt{90} - 4 < |a| < \sqrt{90} + 4.$

Chapter 9

Inequalities

9.1 Some Information

When solving problems, it is often necessary to use the inequalities listed in the following table. The right column of the table indicates the necessary and sufficient condition for which the inequality on the left side of the table becomes equality.

Inequality	Case of equality
$(x - y)^2 \geq 0$	$x = y$
$x^2 + y^2 \geq 2xy$	$x = y$
$x + y \geq 2\sqrt{xy}, \quad x, y \geq 0$	$x = y$
$2^x + 2^y \geq 2 \cdot 2^{(x+y)/2}$	$x = y$
$x + \dfrac{y^2}{x} \geq 2y, \quad x > 0$	$x = y$
$x + \dfrac{y^2}{x} \leq 2y, \quad x < 0$	$x = y$
$\dfrac{2xy}{x^2 + y^2} \leq 1$	$x = y$
$\left(\dfrac{x + y}{2}\right)^2 \leq \dfrac{x^2 + y^2}{2}$	$x = y$
$x^3 + y^3 + z^3 \geq 3xyz, \quad x, y, z > 0$	$x = y = z$

Let's show you how to use this table. For example, the inequality $x^2 + y^2 \geq 2xy$ is valid for all possible values of x and y. An equal sign is executed if and only if $x = y$. If x and y are such that $x \neq y$, then the strict inequality $x^2 + y^2 > 2xy$ holds.

Proofs.[1]

1. Inequalities from the second to the seventh obviously follow from the first inequality, the justifications of which are obvious.
2. The proof of the inequality

$$x^3 + y^3 + z^3 \geq 3xyz, \quad x, y, z > 0,$$

is based on the representation

$$x^3 + y^3 + z^3 - 3xyz = \frac{1}{2} \cdot (x + y + z) \cdot \left((x - y)^2 + (y - z)^2 + (x - z)^2\right).$$

Moreover, as can be seen from this representation, the equality sign in the initial inequality can be achieved only in the case of $x = y = z$.

Here are some more useful inequalities containing a modulus.

Inequality	Case of equality
$\|x\| - x \geq 0$	$x \geq 0$
$\|x\| + x \geq 0$	$x \leq 0$
$\|x\| + \|1 - x\| \geq 1$	$x \in [0; 1]$

3. The validity of the first two inequalities containing the modulus is obvious. Let us prove the inequality $|x| + |1 - x| \geq 1$, in which

---

[1] Since not all of these inequalities are listed in school textbooks, we give their proofs and advise you to reproduce these proofs when solving exam or Olympiad problems.

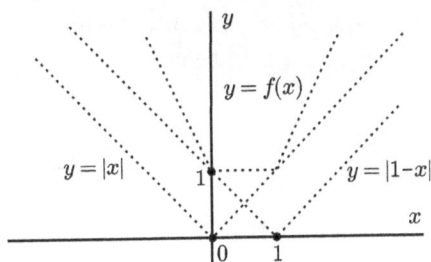

Fig. 9.1. Graph $f(x) = |x| + |1 - x|$.

the equal sign is reached only for $x \in [0; 1]$. Consider the function $f(x) = |x| + |1 - x|$. For $f(x)$, the following representation is valid:

$$f(x) = \begin{cases} 1 - 2x, & x \in (-\infty; 0), \\ 1, & x \in [0; 1], \\ 2x - 1, & x \in (1; +\infty). \end{cases}$$

On the interval $(-\infty; 0)$, the function $f(x)$ (see Fig. 9.1) decreases monotonically; hence, $f(x) > f(0) = 1$, for $x \in (-\infty; 0)$. On the interval $(0; +\infty)$, the function $f(x)$ increases monotonously; therefore, $f(x) > f(1) = 1$, for $x \in (0; +\infty)$.

Also, note another useful inequality, from which we later obtain both the inequality between the arithmetic and geometric means, and the well-known Cauchy–Bunyakovsky inequality.

Theorem 9.1 (Jensen's inequality). *Let f be convex by $[a, b]$ and $x_1, \ldots, x_n \in [a, b]$. Then, for numbers $\alpha_1, \ldots, \alpha_n \geq 0$: $\quad \alpha_1 + \cdots + \alpha_n = 1$ hold true*

$$f(\alpha_1 x_1 + \cdots + \alpha_n x_n) \leq \alpha_1 f(x_1) + \cdots + \alpha_n f(x_n).$$

Proof. We prove it using the method of mathematical induction:

- for the minimum natural number $n = 2$ (obviously, because it coincides with the definition);
- assume that the statement is true for the number $n - 1$;
- we prove for the following natural number n.

Let's put $\beta = \alpha_2 + \cdots + \alpha_n$. Then, $\alpha_1, \beta \geq 0$ (the case of $\beta = 0$ is trivial, so we assume that $\beta > 0$), $\alpha_1 + \beta = 1$ and

$$f\left(\alpha_1 x_1 + \beta\left(\frac{\alpha_2}{\beta}x_2 + \cdots + \frac{\alpha_n}{\beta}x_n\right)\right)$$

$$\leq \alpha_1 f(x_1) + \beta f\left(\frac{\alpha_2}{\beta}x_2 + \cdots + \frac{\alpha_n}{\beta}x_n\right)$$

$$\leq \alpha_1 f(x_1) + \beta\left(\frac{\alpha_2}{\beta}f(x_2) + \cdots + \frac{\alpha_n}{\beta}f(x_n)\right).$$

Next, it remains to use induction for the numbers

$$\frac{\alpha_2}{\beta}, \ldots, \frac{\alpha_n}{\beta} \geq 0; \quad \frac{\alpha_2}{\beta} + \cdots + \frac{\alpha_n}{\beta} = 1. \qquad \square$$

Remark 9.1. If the function is concave, the inequality sign changes:

$$f(\alpha_1 x_1 + \cdots + \alpha_n x_n) \geq \alpha_1 f(x_1) + \cdots + \alpha_n f(x_n).$$

Theorem 9.2. *Let $n \in \mathbb{N}$. Then, for positive x_k, $k = 1, \ldots, n$, the following is true:*

$$\frac{x_1 + \cdots + x_n}{n} \geq \sqrt[n]{x_1 \ldots x_n}.$$

Proof. Consider $f(x) = \ln x$. The function $\ln x$ is convex upward. Let's write Jensen's inequality:

$$\ln(\alpha_1 x_1 + \cdots + \alpha_n x_n) \geq \alpha_1 \ln x_1 + \cdots + \alpha_n \ln x_n,$$

where $\alpha_1, \ldots, \alpha_n \geq 0$, $\alpha_1 + \cdots + \alpha_n = 1$. This is equivalent to

$$\alpha_1 x_1 + \cdots + \alpha_n x_n \geq x_1^{\alpha_1} \ldots x_n^{\alpha_n}.$$

Let's put $\alpha_1 = \cdots = \alpha_n = \frac{1}{n}$. We obtain an inequality connecting the arithmetic mean with the geometric mean:

$$\frac{x_1 + \cdots + x_n}{n} \geq \sqrt[n]{x_1 \ldots x_n}. \qquad \square$$

Theorem 9.3 (Helder's inequality). *Let $1/p + 1/q = 1$, where $p \neq 0, 1$, $q \neq 0, 1$. Then, for valid a_k, $b_k \geq 0$, $k = 1, \ldots, n$, the following is true:*

$$\sum_{k=1}^{n} a_k b_k \leq \left(\sum_{k=1}^{n} a_k^p\right)^{\frac{1}{p}} \left(\sum_{k=1}^{n} b_k^q\right)^{\frac{1}{q}}, \quad p > 1,$$

$$\sum_{k=1}^{n} a_k b_k \geq \left(\sum_{k=1}^{n} a_k^p\right)^{\frac{1}{p}} \left(\sum_{k=1}^{n} b_k^q\right)^{\frac{1}{q}}, \quad p \in (-\infty; 0) \cup (0; 1).$$

Proof. Consider the function $f(x) = x^p$, for $p > 1$. For $p > 1$, this function is convex, so for any $\alpha_1, \ldots, \alpha_n \geq 0$, $\alpha_1 + \cdots + \alpha_n = 1$ completed

$$\left(\sum_{k=1}^{n} \alpha_k x_k\right)^p \leq \sum_{k=1}^{n} \alpha_k (x_k)^p.$$

Let $\frac{1}{p} + \frac{1}{q} = 1$ and $B = \sum_{k=1}^{n} b_k^q$. For simplicity, we assume that $a_k, b_k \geq 0$ because otherwise we consider their expressions modulo. Let us put

$$\alpha_k = \frac{b_k^q}{B}, \quad x_k = \frac{a_k B}{b_k^{q-1}}.$$

Then

$$\left(\sum_{k=1}^{n} a_k b_k\right) \leq \left(\sum_{k=1}^{n} \frac{a_k^p}{B} B^p\right)^{\frac{1}{p}} = \left(\sum_{k=1}^{n} a_k^p\right)^{\frac{1}{p}} \left(\sum_{k=1}^{n} b_k^q\right)^{\frac{1}{q}}.$$

For $p < 1$, the function $f(x) = x^p$ is concave. So, we similarly obtain Helder's inequality for $p < 1$, $p \neq 0$, with the inverse sign in the inequality. $\square$

Remark 9.2. Let $\vec{a}, \vec{b} \in \mathbb{R}^d$. Then,

$$\vec{a} = (a_1, \ldots, a_d) \quad \vec{b} = (b_1, \ldots, b_d).$$

Let's put

$$\|\vec{a}\|_p = \left(\sum_{k=1}^{d} a_k^p\right)^{\frac{1}{p}}.$$

Then, Helder's inequality is equivalent to the following:

$$(\vec{a}, \vec{b}) \leq \|\vec{a}\|_p \cdot \|\vec{b}\|_q, \quad \frac{1}{p} + \frac{1}{q} = 1, \quad p > 1.$$

In the case of $p = q = 2$,

$$(\vec{a}, \vec{b}) \leq \|\vec{a}\|_2 \cdot \|\vec{b}\|_2. \tag{9.1}$$

This last inequality explains the correctness of introducing an angle between the vectors $\vec{a}$ and $\vec{b}$ using the following formula:

$$\cos(\widehat{\vec{a}, \vec{b}}) = \frac{(\vec{a}, \vec{b})}{\|\vec{a}\|_2 \cdot \|\vec{b}\|_2}.$$

Also, inequality (9.1) is well-known as Cauchy–Bunyakovsky inequality.

9.2   Solving Problems from Exams and Olympiads

Example 9.1. Solve the equation

$$\frac{25}{\sqrt{x-1}} + \frac{4}{\sqrt{y-2}} = 14 - \sqrt{x-1} - \sqrt{y-2}.$$

Proof. Let us write the original equation in the form

$$\left(\sqrt{x-1} + \frac{25}{\sqrt{x-1}}\right) + \left(\sqrt{y-2} + \frac{4}{\sqrt{y-2}}\right) = 14.$$

Note that the inequality $t + y^2/t \geq 2 \cdot y$, $t > 0$, implies the inequalities

$$\sqrt{x-1} + \frac{25}{\sqrt{x-1}} \geq 2 \cdot 5 = 10,$$

$$\sqrt{y-2} + \frac{4}{\sqrt{y-2}} \geq 2 \cdot 2 = 4;$$

therefore,

$$\left(\sqrt{x-1} + \frac{25}{\sqrt{x-1}}\right) + \left(\sqrt{y-2} + \frac{4}{\sqrt{y-2}}\right) \geq 14.$$

If the sum of two terms, the first of which is not less than 10 and the second is not less than 4, is 14, then the first term is 10, and the second is 4.

Let's use the fifth row of the table. Since the equal sign in an inequality of the form $t + y^2/t \geq 2 \cdot y$, $t > 0$, is achieved only in the case of $t = y$, the original equation is equivalent to the system of equations

$$\begin{cases} \sqrt{x-1} = 5, \\ \sqrt{y-2} = 2. \end{cases} \iff \begin{cases} x = 26, \\ y = 6. \end{cases}$$

Answer: $(26; 6)$. □

Example 9.2. For what values of a and b does the inequality

$$b < 16^{\frac{2x-1}{4x^2-4x+5}} \leqslant a$$

hold for all real x?

Proof. Let $t = 2x - 1$. Consider the function $f(t) = \frac{t}{t^2+4}$. Since

$$t^2 + 4 \geqslant 2\sqrt{4t^2} = 4|t|,$$

then

$$-\frac{1}{4} \leqslant \frac{t}{t^2 + 4} \leqslant \frac{1}{4}.$$

The values of $\pm\frac{1}{4}$ are reached at $t = \pm 2$. Hence, the set of values of the functions $g(t) = 16^{\frac{2x-1}{4x^2-4x+5}} = 16^{\frac{t}{t^2+4}}$ is a segment $[\frac{1}{2}; 2]$.

Answer: $a \geq 2$, $b < \frac{1}{2}$. □

Example 9.3. Solve the equation

$$2^{2^{\sin^2 x}} + 2^{2^{(\cos 2x)/2}} = 2^{1+\sqrt[4]{2}}.$$

Proof. Recall the inequality $2^a + 2^b \geq 2 \cdot 2^{(a+b)/2}$, which is true for all a and b. For this, equality is achieved only in the case of $a = b$. Applying this inequality twice to the original problem, we obtain

$$2^{2^{\sin^2 x}} + 2^{2^{(\cos 2x)/2}} = 2^{2^{\sin^2 x}} + 2^{2^{1/2 - \sin^2 x}} \geq 2 \cdot 2^{(2^{\sin^2 x} + 2^{1/2 - \sin^2 x})/2}$$

$$= 2^{1 + (1/2)\left(2^{\sin^2 x} + 2^{(1/2) - \sin^2 x}\right)}$$

$$\geq 2^{1 + 2^{(\sin^2 x + (1/2) - \sin^2 x)/2}} = 2^{1 + \sqrt[4]{2}}.$$

However, according to the original equation, it is required that all inequalities be equalities, which means

$$\sin^2 x = \frac{1}{2} - \sin^2 x.$$

Solving the equation, we find $\sin x = \pm 1/2$, and $x = \pm \pi/6 + \pi n$, $n \in \mathbb{Z}$.

Answer: $\pm \pi/6 + \pi n$, $n \in \mathbb{Z}$. ▫

Example 9.4. The positive numbers a, b, and c satisfy the relation

$$a^2 + b^2 + c^2 = 1.$$

Find the largest possible value of the expression $ab + bc\sqrt{3}$.

Proof. Here are two solutions:

1. Let's make a substitution: $a = \cos \varphi \cos \theta$, $c = \sin \varphi \cos \theta$, $b = \sin \theta$. This substitution is well known as the spherical coordinate system, and it is based on the identity $\cos^2 \alpha + \sin^2 \alpha = 1$. Since the above substitution leads the equation $a^2 + b^2 + c^2 = 1$ to the identity $1 = 1$, we get

$$ab + bc\sqrt{3} = \sin \theta \cos \theta (\cos \varphi + \sqrt{3} \sin \varphi) = \sin 2\theta \sin \left(\varphi + \frac{\pi}{6}\right) \leq 1.$$

   Moreover, the equal sign is achieved when $\theta = \pi/4$ and $\varphi = \pi/3$. That is, when $a = \frac{1}{2\sqrt{2}}$, $c = \frac{\sqrt{3}}{2\sqrt{2}}$, and $b = \frac{1}{\sqrt{2}}$.

2. Let us use the Cauchy–Bunyakovsky inequality (9.1)

$$b(a + c\sqrt{3}) \leq b\sqrt{a^2 + c^2}\sqrt{1 + 3} = 2b\sqrt{a^2 + c^2} \leq b^2 + a^2 + c^2 = 1.$$

   Equality is achieved when $a = c\sqrt{3}$ and $b^2 = a^2 + c^2$ are fulfilled, i.e., on the same numbers when $a = \frac{1}{2\sqrt{2}}$, $c = \frac{\sqrt{3}}{2\sqrt{2}}$, and $b = \frac{1}{\sqrt{2}}$.

Answer: 1. ▫

Example 9.5. Find all values of $a > 0$ for which there are positive solutions to the inequality

$$\frac{x^3}{a + 2013^{4/3}x} + \frac{2013^{4/3}x}{a + x^3} \le \frac{3}{2} - \frac{a}{x(x^2 + 2013^{4/3})}.$$

Proof. Let us prove one auxiliary lemma.

Lemma 9.1. *For a, b, $c > 0$, the following inequality holds:*

$$\frac{a}{b + c} + \frac{b}{a + c} + \frac{c}{a + b} \ge \frac{3}{2}.$$

Moreover, the equal sign is achieved if and only if $a = b = c$.

Proof. There are many ways to prove this inequality. Here is a method based on the substitution of a variable. We introduce the notation $u = a+b$, $v = a + c$, and $w = b + c$. Note that $u, v, w > 0$ and

$$2a = u + v - w, \quad 2b = u - v + w, \quad 2c = -u + v + w.$$

With the new variables, the original inequality is equivalent to

$$\frac{u + v - w}{w} + \frac{u - v + w}{v} + \frac{-u + v + w}{u} \ge 3$$

$$\Longleftrightarrow \frac{u + v}{w} + \frac{u + w}{v} + \frac{v + w}{u} \ge 6$$

$$\Longleftrightarrow \left(\frac{u}{w} + \frac{w}{u}\right) + \left(\frac{u}{v} + \frac{v}{u}\right) + \left(\frac{v}{w} + \frac{w}{v}\right) \ge 6.$$

The last inequality follows from the estimate $t + 1/t \ge 2$, and the equal sign is achieved if and only if $t = 1$. For the last outlier inequality, this means that the equal sign in it is achieved if and only if $u = v = w$. However, the condition $u = v = w$ is equivalent to the condition $a = b = c$. $\square$

If we substitute $b = x^3$, $c = 2013^{4/3}x$ into the proved inequality, then we get the inequality

$$\frac{x^3}{a + 2013^{4/3}x} + \frac{2013^{4/3}x}{a + x^3} + \frac{a}{x(x^2 + 2013^{4/3})} \ge \frac{3}{2}.$$

According to the task condition,

$$\frac{x^3}{a + 2013^{4/3}x} + \frac{2013^{4/3}x}{a + x^3} + \frac{a}{x(x^2 + 2013^{4/3})} \le \frac{3}{2}.$$

Thus, equality arises. It follows from the lemma that the evaluation is achieved only with the following equality:

$$a = x^3 = 2013^{4/3}x.$$

Having solved the last equations, we find $x = 2013^{2/3}$, $a = 2013^2$. From here, with $a = 2013^2$, the solution is $x = 2013^{2/3}$. There are no positive solutions for other $a > 0$.

Answer: 2013^2. $\square$

Example 9.6. For what values of a does the system

$$\begin{cases} |x - a| + |y - a| + |a + 1 - x| + |a + 1 - y| = 2, \\ y + 2|x - 5| = 6 \end{cases}$$

have a single solution?

Proof. By rearranging and using the inequality $|t| + |1 - t| \geq 1$, we can conclude that for the left side in the first equation of the system, the following inequality is true:

$$|x - a| + |y - a| + |a + 1 - x| + |a + 1 - y|$$
$$= (|x - a| + |1 - (x - a)|) + (|y - a| + |1 - (y - a)|) \geq 2.$$

However, since the condition of the problem states that this sum is equal to 2, it follows from the last inequality in the second table that each of the terms in parentheses is equal to 1, and since the equal sign is achieved only for $t \in [0; 1]$, the original system is equivalent to the following system:

$$\begin{cases} |x - a| + |1 - (x - a)| = 1, \\ |y - a| + |1 - (y - a)| = 1, \iff \\ y + 2|x - 5| = 6. \end{cases} \begin{cases} 0 \leq x - a \leq 1, \\ 0 \leq y - a \leq 1, \\ y + 2|x - 5| = 6. \end{cases}$$

$$\iff \begin{cases} a \leq x \leq 1 + a, \\ a \leq y \leq 1 + a, \\ y + 2|x - 5| = 6. \end{cases}$$

Since $y = 6 - 2|x - 5|$, the system has a unique solution if the only solution x of the system is

$$\begin{cases} a \le x \le 1 + a, \\ a \le 6 - 2|x - 5| \le 1 + a. \end{cases} \iff \begin{cases} a \le x \le 1 + a, \\ 5 - a \le 2|x - 5| \le 6 - a. \end{cases}$$

Consider the following cases:

1. Let $a > 6$. Then, there are no solutions.
2. Let $a \in [5; 6]$. Then, the system is equivalent to the following:

$$\begin{cases} a \le x \le 1 + a, \\ \dfrac{a - 6}{2} \le x - 5 \le \dfrac{6 - a}{2}. \end{cases} \iff \begin{cases} a \le x \le 1 + a, \\ \dfrac{a + 4}{2} \le x \le \dfrac{16 - a}{2}. \end{cases}$$

The latter system has a single solution (see Figs. 9.2 and 9.3) if $a = (16 - a)/2$, i.e., $a = 16/3 \in [5; 6]$, or if $1 + a = (a + 4)/2$, i.e., $a = 2 \notin [5; 6]$. Therefore, it suits us that $a = 16/3$.
3. Let $a < 5$. Then, the system takes the form

$$\begin{cases} a - 5 \le x - 5 \le a - 4, \\ \dfrac{5 - a}{2} \le |x - 5| \le \dfrac{6 - a}{2}. \end{cases} \iff \begin{cases} a - 5 \le t \le a - 4, \\ \dfrac{5 - a}{2} \le |t| \le \dfrac{6 - a}{2}, \end{cases}$$

where $t = x - 5$. Since $a - 5 < (a - 5)/2$, the latter system has a unique solution only in the case of (see Fig. 9.4) $a - 4 = (a - 6)/2$, i.e., $a = 2$.

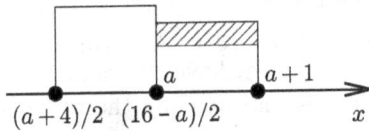

Fig. 9.2.   Case II.

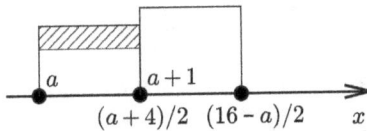

Fig. 9.3.   Case II.

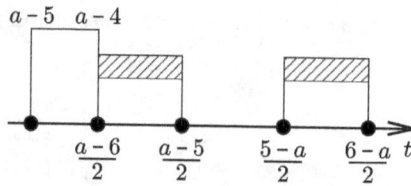

Fig. 9.4.   Case III.

Since $a = 2$ satisfies the condition $a < 5$, then for $a = 2$, the original system has a unique solution.

Answer: $a = 2$, $a = 16/3$. $\square$

Example 9.7. At the base of the pyramid $SABCD$ is a trapezoid $ABCD$ with bases BC and AD. The points P_1, P_2, and P_3 belong to the side of BC, and $BP_1 < BP_2 < BP_3 < BC$. The points Q_1, Q_2, and Q_3 belong to the side of AD, and $AQ_1 < AQ_2 < AQ_3 < AD$. Let's denote the intersection points of BQ_1 with AP_1, P_2Q_1 with P_1Q_2, P_3Q_2 with P_2Q_3, and CQ_3 with P_3D through R_1, R_2, R_3, and R_4, respectively. It is known that the sum of the volumes of the pyramids $SR_1P_1R_2Q_1$ and $SR_3P_3R_4Q_3$ is equal to 78. Find the minimum value of the expression

$$V_{SABR_1}^2 + V_{SR_2P_2R_3Q_2}^2 + V_{SCDR_4}^2.$$

Proof. From the properties of the trapezoid, it follows that the triangles (see Fig. 9.5) $S_{\triangle ABR_1} = S_{\triangle R_1P_1Q_1}$, $S_{\triangle P_1Q_1R_2} = S_{\triangle R_2P_2Q_2}$, $S_{\triangle P_2Q_2R_3} = S_{\triangle R_3P_3Q_3}$, $S_{\triangle P_3Q_3R_4} = S_{\triangle R_4CD}$ have the same area. Hence, the equality of the sums of the areas is indicated by the same color in Fig. 9.6. The equality of the sums of the areas takes the form

$$S_{ABR_1} + S_{R_2P_2R_3Q_2} + S_{CDR_4} = S_{R_1P_1R_2Q_1} + S_{R_3P_3R_4Q_3}.$$

Therefore,

$$V_{SABR_1} + V_{SR_2P_2R_3Q_2} + V_{SCDR_4} = V_{SR_1P_1R_2Q_1} + V_{SR_3P_3R_4Q_3} = 78.$$

Let's put $a_1 = V_{SABR_1}$, $a_2 = V_{SR_2P_2R_3Q_2}$, $a_3 = V_{SCDR_4}$. From the problem condition, $a_1 + a_2 + a_3 = 78$, we are looking for $a_1^2 + a_2^2 + a_3^2 \to \min$,

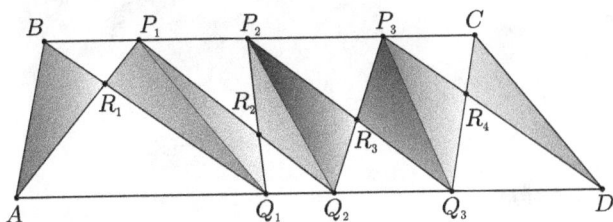

Fig. 9.5. Trapezoids.

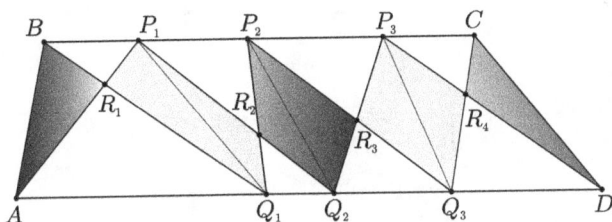

Fig. 9.6. Equal areas.

provided that a_1, a_2, a_3 are non-negative. Inequality is fair:

$$\left(\frac{a_1}{3} + \frac{a_2}{3} + \frac{a_3}{3}\right)^2 \le \frac{a_1^2 + a_2^2 + a_3^2}{3}.$$

This inequality is justified by applying the inequality $2ab \le a^2 + b^2$ three times or by using the convexity of the function $f(x) = x^2$. We get

$$a_1^2 + a_2^2 + a_3^2 \ge \frac{(a_1 + a_2 + a_3)^2}{3} = 2028.$$

The sign of equality in the inequality is achieved when $a_1 = a_2 = a_3 = 26$.

Answer: 2028. □

Problems

9.1. At what values of p is the equation

$$4^x + 2^{x+2} + 7 = p - 4^{-x} - 2 \cdot 2^{1-x}$$

has a solution?

9.2. Find all the solutions of the system

$$\begin{cases} xy - t^2 = 9, \\ x^2 + y^2 + z^2 = 18. \end{cases}$$

9.3. Solve the system

$$\begin{cases} \sqrt{x+2} + \sqrt{x^2 + 5x + 5} \geq 2, \\ x^2 + 6x + 5 \leq 0. \end{cases}$$

9.4. Find the largest value of the function

$$\frac{10^x}{25^{x-1} + 10^x + 4^{x+1}}.$$

9.5. Solve the system of equations

$$\begin{cases} \tan^2 x + \cot^2 x = 2\sin^2 y, \\ \sin^2 y + \cos^2 z = 1. \end{cases}$$

9.6. For what values of a and b is the inequality

$$b \leq 9^{\frac{6x+3}{4x^2+4x+10}} < a$$

is executed, for all valid x?

9.7. The positive numbers a, b, c satisfy the relation

$$a^2 + 5b^2 + c^2 = 1.$$

Find the largest possible value of the expression $ab\sqrt{3} + bc$.

9.8. For each value of c, solve the system

$$\begin{cases} \dfrac{9}{\sqrt{x+c}} + \dfrac{16}{\sqrt{y-c}} \leq 22 - \sqrt{x+c} - 4\sqrt{y-c}, \\ 2^{x-11}\log_2(4-y) = 1. \end{cases}$$

9.9. Solve the equation

$$\tan^2(5x + \sin^2 y) + \left| \frac{5x + \cos 2y}{3} + \frac{3}{5x + \cos 2y} \right| = 4\cos^2 \frac{7\pi}{4}.$$

9.10. For each value of a, solve the system

$$\begin{cases} 4\log_4^2 x + 9\log_8^2 y \le 4(a^2 + a), \\ \log_2^2 xy \ge 8(a^2 + a). \end{cases}$$

9.11. Find the largest value of a for which the inequality

$$a\sqrt{a}(x^2 - 2x + 1) + \frac{\sqrt{a}}{x^2 - 2x + 1} \le \sqrt[4]{a^3} \cdot |\sin(\pi x/2)|$$

has at least one solution.

9.12. Find all pairs of numbers $(x; y)$ that satisfy the equation

$$\tan^4 x + \tan^4 y + 2\cot^2 x \cdot \cot^2 y = 3 + \sin^2(x + y).$$

9.13. Find all pairs of positive numbers (x, y) satisfying the equation

$$\log_{2x^2y+1}(x^4 + y^2 + 1) = \log_{y^4+x^2+1}(2xy^2 + 1).$$

9.14. Prove that all solutions of the inequality

$$\sqrt{x-1} + \sqrt[3]{x^2 - 1} > 2$$

satisfy the inequality

$$x + 2\sqrt{x-1} + \sqrt[3]{x^4 - 2x^2 + 1} > 1 + 2\sqrt[3]{x^2 - 1}.$$

9.15. For what values of a does the system

$$\begin{cases} |x + a| + |y - a| + |a + 1 + x| + |a + 1 - y| = 2, \\ y = 2|x - 4| - 5 \end{cases}$$

have a single solution?

9.16. Find all pairs of numbers $(x; y)$ that satisfy the equation

$$\left(\cos^2 x + \frac{1}{\cos^2 x}\right)^2 + \left(\sin^2 x + \frac{1}{\sin^2 x}\right)^2 = +12 + \frac{\sin y}{2}.$$

9.17. Solve the equation

$$2^{2\cos^2 x} + 2^{2-(\cos 2x)/2} = 2^{1+\sqrt[4]{2}}.$$

9.18. Find all pairs of numbers x and y satisfying the system of inequalities

$$\begin{cases} 4^{x+y-1} + 3 \cdot 4^{2y-1} \leq 2, \\ x + 3y \geq 2 - \log_4 3. \end{cases}$$

9.19. Find the largest and smallest values of the expression $x^2 + 2y^2$ if

$$x^2 - xy + 2y^2 = 1.$$

9.20. Find all the values of b for which the inequality

$$(3 - 2\sqrt{2})^x + (b^4 + 12 - 6b^2) \cdot (3 + 2\sqrt{2})^x + 9^t + b^2/4 + b \cdot 3^t - \sqrt{12} \leq 0$$

has at least one solution (t, x).

9.21. For each value of $a \geq 1/(2\pi)$, find all the roots of the equation

$$\cos\left(\frac{2x + a}{2x^2 + 2ax + 5/2a^2}\right) = \cos\left(\frac{2x - a}{2x^2 - 2ax + 5/2a^2}\right).$$

9.22. Find the largest value of ω for which the following system has a solution:

$$\begin{cases} 4\sin^2 y - \omega = 16\sin^2 \dfrac{2x}{7} + 9\cot^2 \dfrac{2x}{7}, \\ (\pi^2 \cos^2 3x - 2\pi^2 - 72)y^2 = 2\pi^2(1 + y^2)\sin 3x. \end{cases}$$

9.23. Find the smallest possible value of the expression

$$\frac{c - b}{a + 2b + c} + \frac{2b}{a + b + 2c} - \frac{4c}{a + b + 3c}$$

with positive a, b, and c.

9.24. Find all the values of a, for each of which the inequality

$$4^x + 4^{-x} + 8/2^x + 2^{-x} - a| + 11a < 26 + 2a(2^x + 2^{-x})$$

has at least one solution.

9.25. Find all values of $a > 0$ for which there are positive solutions to the inequality

$$\frac{x^3}{a + 2014^{4/3}x} + \frac{2014^{4/3}x}{a + x^3} \leq \frac{3}{2} - \frac{a}{x(x^2 + 2014^{4/3})}.$$

9.26. A parcel must be packed in a box in the form of a rectangular parallelepiped and tied once lengthwise and twice across (see Fig. 9.7). Is it possible to send a parcel of 37 dm^3 tied with 3.6 m of rope (the thickness of the walls of the box and the rope up to the nodes are neglected)?

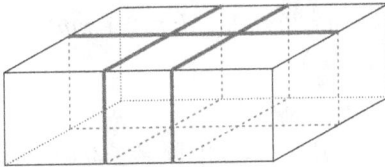

Fig. 9.7.   Parcel.

9.27. At the base of the pyramid $SABCD$ is a trapezoid $ABCD$ with bases BC and AD. The points P_1, P_2, and P_3 belong to the side of BC, and $BP_1 < BP_2 < BP_3 < BC$. The points Q_1, Q_2, and Q_3 belong to the side of AD, and $AQ_1 < AQ_2 < AQ_3 < AD$. Let us denote the intersection points of BQ_1 with AP_1, P_2Q_1 with P_1Q_2, P_3Q_2 with P_2Q_3, and CQ_3 with P_3D through R_1, R_2, R_3, and R_4, respectively. It is known that the sum of the volumes of the pyramids $SR_1P_1R_2Q_1$ and $SR_3P_3R_4Q_3$ is 96. Find the minimum value of the expression

$$V_{SABR_1}^2 + V_{SR_2P_2R_3Q_2}^2 + V_{SCDR_4}^2.$$

Hints and Answers

9.1. $p \in [17; +\infty)$.

9.2. $(x, y, t, z) = (3, 3, 0, 0), (-3, -3, 0, 0)$.

9.3. -1.

9.4. $5/9$.

9.5. $(x, y, z) = (\pi/4 + \pi k/2; \pi/2 + \pi l; \pi/2 + \pi m)$, $k, l, m \in \mathbb{Z}$.

9.6. $a > 3$, $b \leq \frac{1}{3}$.

9.7. $1/\sqrt{5}$.

9.8. If $c = -2$, then $(x; y) = (11; 2)$; if $c \neq -2$, then there are no solutions.

9.9. $(x; y) = ((2\pi + 2)/15; \pm \arcsin \sqrt{(\pi - 2)/3} + \pi k), (-(2\pi + 4)/15; \pm \arcsin \sqrt{(4 - \pi)/3} + \pi m)$, $k, m \in \mathbb{Z}$.

9.10. If $a \in (-1; 0)$, then there are no solutions; if $a = -1$, $a = 0$, then $x = y = 1$; if $a \in (-\infty; -1) \cup (0; +\infty)$, then $x_1 = y_1 = 2^{\sqrt{2(a^2+a)}}$, $x_2 = y_2 = 2^{-\sqrt{2(a^2+a)}}$.

9.11. $a = 1/16$.

9.12. $x = \pi/4 + \pi n$, $y = \pi/4 + \pi k$, $n, k \in \mathbb{Z}$; $x = -\pi/4 + \pi n$, $y = -\pi/4 + \pi k$, $n, k \in \mathbb{Z}$.

9.13. $x = y = 1$.

9.15. $a = -2$; $a = -16/3$. **Hint:** Use the inequality $|z| + |1 - z| \geq 1$.

9.16. $x = \pi/4 + \pi n/2$, $y = \pi/2 + 2\pi k$, $n, k \in \mathbb{Z}$.

9.17. $\pm \pi/3 + \pi n$, $n \in \mathbb{Z}$.

9.18. $x = 1/2 + (1/2) \log_4 3$, $y = 1/2 - (1/2) \log_4 3$.

9.19. $(2\sqrt{2}/(2\sqrt{2} - 1)), (2\sqrt{2}/(2\sqrt{2} + 1))$.

9.20. $b = -\sqrt{3}$.

9.21. $x = 0$; $x = \sqrt{5}a/2$; $x = -\sqrt{5}a/2$.

9.22. -14.

9.23. $6\sqrt{2} - 9$.

9.24. $a \in (-8; 4) \cup (7; +\infty)$.

9.25. 2014^2.

9.26. No.

9.27. 3072.

Index

www.ingramcontent.com/pod-product-compliance
Lightning Source LLC
Chambersburg PA
CBHW061627220326
41598CB00026BA/3911